Protein and Peptide-based Microarrays for Multiplex Detection

Protein and Peptide-based Microarrays for Multiplex Detection

Edited by
Navid Rabiee
Sharif University of Technology, Tehran, Iran

Michael R Hamblin
Wellman Center for Photomedicine, Massachusetts General Hospital, Boston, MA, USA

IOP Publishing, Bristol, UK

ISBN 978-0-7503-3667-3 (ebook)
ISBN 978-0-7503-3665-9 (print)
ISBN 978-0-7503-3668-0 (myPrint)
ISBN 978-0-7503-3666-6 (mobi)

DOI 10.1088/978-0-7503-3667-3

Version: 20201001

IOP ebooks

British Library Cataloguing-in-Publication Data: A catalogue record for this book is available from the British Library.

Published by IOP Publishing, wholly owned by The Institute of Physics, London

IOP Publishing, Temple Circus, Temple Way, Bristol, BS1 6HG, UK

US Office: IOP Publishing, Inc., 190 North Independence Mall West, Suite 601, Philadelphia, PA 19106, USA

Contents

Preface

Microarray technology allows the simultaneous analysis of several parameters within a single experiment. Microspots of capture molecules are immobilized in rows and columns onto a solid support and exposed to samples containing the corresponding binding molecules. Readout systems based on fluorescence, chemiluminescence, mass spectrometry, radioactivity or electrochemistry can be used to detect complex formation within each microspot. Such miniaturized and parallelized binding assays can be highly sensitive, and the extraordinary power of the method is exemplified by array-based gene expression analysis. In these systems, arrays containing immobilized DNA probes are exposed to complementary targets and the degree of hybridization is measured. Recent developments in the field of protein microarrays show applications for enzyme–substrate, DNA–protein and different types of protein–protein interactions. In this book, we comprehensively discuss several aspects of protein and peptide-based microarrays for multiplex detection based on Rabiee's theory (variable laws) which we explained in the book entitled '*Introduction to Nanomaterials in Medicine*' by Rabiee *et al* for the first time.

Acknowledgements

Firstly, the authors would like to express their gratitude to all who helped them. Secondly, the authors would like to express their heartfelt gratitude to their beloved families for all their love and encouragement through the years towards to completion of this book; their parents who raised them with a love for science and a conscience; and also their wives. Finally, it is our pleasure to acknowledge the guidance and contribution of the Production team, at IOP Publishing, for their expert help.

Editor biographies

Navid Rabiee

Navid Rabiee is a PhD candidate of inorganic chemistry at Sharif University of Technology, Tehran, Iran, under supervision of Professor M Bagherzadeh. Since 2016, he has worked on several fields with a multidisciplinary approach by blending modern molecular and cellular biology/biochemical sciences with engineering principles to design the next-generation of medical systems and devices for patient treatment. His work has resulted in publication of over 80 peer-reviewed journal articles in prestigious journals including *Nature*, *Nature Medicine*, *The Lancet*, *Nano Today* etc, 5 books, over 20 chapters and 5 conference papers.

Michael R Hamblin

Michael R Hamblin PhD is a Principal Investigator at the Wellman Center for Photomedicine, Massachusetts General Hospital, and an Associate Professor at Harvard Medical School. He has interests in photodynamic therapy (PDT) and photobiomodulation. He has published 420 peer-reviewed articles, is Editor-in-Chief of *Photobiomodulation*, *Photomedicine* and *Laser Surgery*, Associate Editor for 10 other journals and serves on NIH Study-Sections. He has an h-factor of 91 and >33 000 citations. He has authored/edited 23 textbooks on PDT and photomedicine including SPIE proceedings. Dr Hamblin was elected as a Fellow of SPIE in 2011, received the first Endre Mester Lifetime Achievement Award Photomedicine from NAALT in 2017, and an Outstanding Career Award from the Dose Response Society in 2018.

List of contributors

Sakineh Hajebi

Sakineh Hajebi graduated with an MSc degree in polymer engineering from Sahand University of technology, Tabriz, Iran, in 2017. During her MSc research, she worked on polymerization and responsive polymers under the supervision of Professor Hossein Roghani-Mamaqani. During this period, she focused on drug delivery systems based on biocompatible and biodegradable polymers. In 2018, she joined Advanced Technologies Research Group in association with Professor Mohammad Rabiee and Professor Michael R Hamblin, working on smart nanostructures applied in therapeutic agent delivery systems employed for diagnosis and therapy of various diseases and disorders. Her work has resulted in the publication of three peer-reviewed journal articles, and three conference papers.

Sepideh Ahmadi

Sepideh Ahmadi received her MSc in molecular genetics from Zabol University, Iran. During her MSc research, she worked on specific detection of Klebsiella pneumoniae by gold nanoparticles probe. Currently, she is working at Shahid Beheshti University of Medical Sciences as a PhD candidate in the field of molecular medicine. Her current research interests include the design of nanoparticles in drug/gene delivery and gene manipulation.

Mohammad Rabiee

Mohammad Rabiee, PhD, is an associate professor of biomedical engineering at the Amirkabir University of Technology, Tehran, Iran. His current research interests include smart drug delivery systems, tissue engineering and biological sensors. He has published over 100 ISI papers and also over 70 International Conference papers; in addition, he has been teaching and performed research for 27 years at Amirkabir University of Technology, Tehran, Iran.

Mojtaba Bagherzadeh

Mojtaba Bagherzadeh, PhD, is a full professor of Inorganic Chemistry at the Sharif University of Technology, Tehran, Iran. His current research interests include inorganic chemistry, inorganic catalysis and bioinorganic chemistry. He has published over 100 ISI papers, with the h-Index of 32.

Saeid Maghsoudi

Saeid Maghsoudi is a master graduate of medicinal chemistry from Shiraz University of Technology, Shiraz, Iran. During his MSc research, he has worked on medicinal and organic chemistry research, particularly, synthesis of bioactive compounds, heterocycle chemistry and natural products. His current research interests include, polymer blends, smart drug delivery systems, aptamers as well as spectroscopy analysis. His work has resulted in the publication of three well-stablished journal articles. Additionally, he is working in a pharmaceutical laboratory at one of the thriving companies in Tehran, Iran.

IOP Publishing

Navid Rabiee and Michael R Hamblin

Chapter 1

Introduction to protein microarrays

Navid Rabiee, Sakineh Hajebi, Sepideh Ahmadi, Mohammad Rabiee, Mojtaba Bagherzadeh, Saeid Maghsoudi and Michael R Hamblin

1.1 Background, history and development

In 1975, microarray technology was first described by Ed Southern and coworkers. They described the use of DNA microarrays for detecting the simultaneous expression of thousands of genes in a wide range of genomic samples. The microarrays used hybridization to identify a specific DNA sequence within a sample, in a similar manner to Southern blots, which consist of immobilization of DNA molecules. Next, complementary DNA or oligonucleotide sequences are applied to hybridize with the DNA molecules to be analyzed [1–5]. In 1984 Geysen *et al* introduced multipin technology for the synthesis of peptide libraries. This was the beginning of combinatorial synthesis of peptides [6]. Therefore, the first combinatorial library method was devised in the form of a chemical array. Libraries of peptides, each of which was pure and specific, could be synthesized in 96 well plates, and they were used for epitope-mapping of B-cells. Nevertheless, this technique was only really useful for low-throughput applications and produced only a small total number of peptides. A few years later, this method was modified by synthesizing and analyzing multiple peptides as spots on cellulose paper (the so-called SPOT library method) by Frank *et al* [7]. The technique of enzyme-linked calorimetry was used to screen the array and then differently colored beads were separated to determine the sequence.

Fodor *et al* described a parallel chemical synthesis technique that was spatially addressable with the use of the light-directed coupling reactions to produce peptide microarrays in 1991 [8]. They were able to identify the successful sequences by attaching a fluorescently labeled antibody. Ironically, the DNA microarray was the first to benefit from Fodor's idea, rather than the peptide array. The advance of this innovative technology was slower in the case of peptide arrays because of the lower quality of peptides synthesized, due to poor performance in peptide coupling and

doi:10.1088/978-0-7503-3667-3ch1

deprotection periods on the glass surface. Nucleic acid arrays progressed to the fabrication of oligonucleotide-based arrays, which afterwards led to the commercialization of the GeneChip™ the core technology of Affymetrix [9–13].

Roger Ekins and coworkers (London, UK) published a paper in 1989 describing the first microarray termed 'micro spot multi-analyte immunoassays' [14]. In 1988 at a Swedish symposium [15] Ekins stated 'multi-analyte immunoassay systems of this kind are likely to totally transform medical diagnosis in the foreseeable future and are also likely to be of value in the analysis of complex protein mixtures deriving from recombinant DNA technologies'. This was a quite fanciful statement at that time, but did indeed lead to a real paradigm shift in analytical chemistry. In the early 1990s, photolithographic techniques for the synthesis of high-density arrays of peptides and oligonucleotides were described by Alejandro Zaffaroni, Stephen P A Fodor (Affymax, Inc.) and coworkers (figure 1.1) [8, 16–18].

The term 'microarray' was probably first used by D R Walt (Tufts University, USA) [19] and P O Brown (Stanford University, USA) [20], and their coworkers, and it should be noted that before 1995, the term 'microarray' was largely unknown.

Both public health and military and civilian defense concerns necessitate the development and availability of analytical technologies capable of the ready detection and reliable identification of biological weapons threats (biothreats). For the purposes of this discussion, biothreats may consist of pathogens, viruses, and toxins. Examples of naturally occurring biothreats, include emerging infectious diseases, infections that often appear unexpectedly and have the potential to spread throughout a population, or increase in severity at each incidence. Recent outbreaks include a highly contagious African viral hemorrhagic fever (Ebola) in 1974, hantavirus pulmonary syndrome in the southwestern United States in 1993, severe

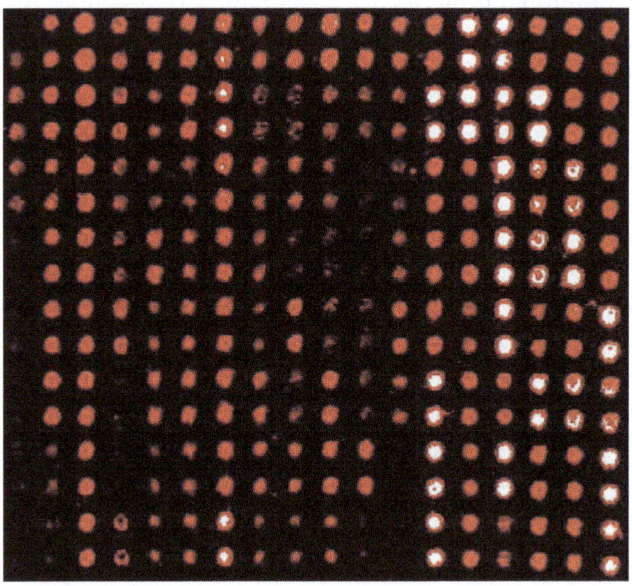

Figure 1.1. Schematic illustration of a fluorescent microarray.

acute respiratory syndrome (SARS) in 2003 [21–24] and the recent emergence of the novel coronavirus (COVID-19). Biothreats with relevance to defense concerns include ricin, anthrax, and botulinum toxin.

1.2 Protein structure

Proteins are biochemical macromolecules taking either globular or fibrous forms, involving one or more polypeptides typically folded into a biologically functional tertiary structure [25–28]. A polypeptide is a single linear polymeric chain of amino acids with peptide bonds between the amino and carboxyl groups of adjacent amino acid residues. The sequence of amino acids in a protein is governed by the sequence of nucleic acids in a gene encoded by the genetic code. Generally, the genetic code assigns codons to 20 standard amino acids; although, in some organisms, the genetic code can include selenocysteine—and in certain archaea—pyrrolysine as amino acids. The residues in a protein are often changed by post-translational modification either immediately or even during synthesis. This involves folding, stability, physical and chemical properties, activity, and protein function. Sometimes proteins contain non-peptide groups attached to certain amino acids, which are called prosthetic groups or cofactors. Proteins can work together to carry out specific functions, and they often associate with each other to form stable multi-protein complexes [29–34].

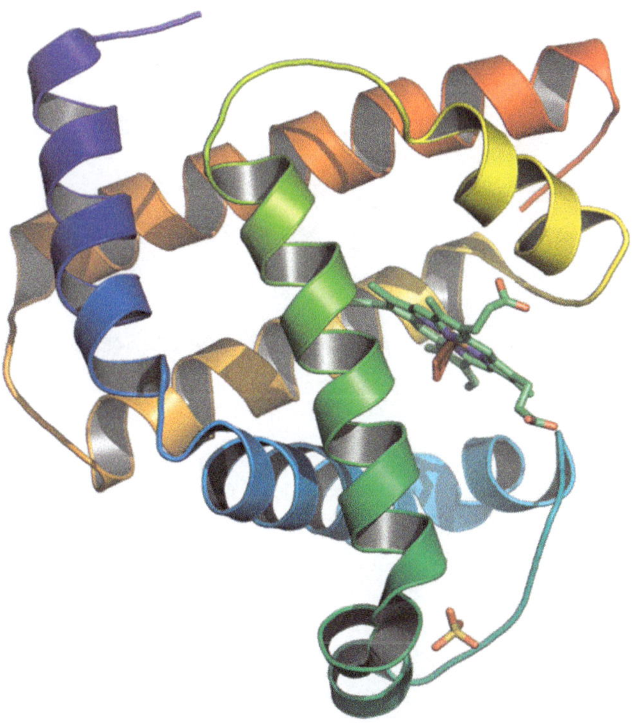

Figure 1.2. A typical structure of a protein.

The ability of polypeptides to self-assemble into a globular state or 'tertiary structure' is one of the most characteristic features of proteins. The degree to which proteins fold into a certain structure can be widely different. Some proteins are considered to have a single structure, because they fold into a highly rigid structure with only small fluctuations. Other proteins can shift from one conformation to another, and undergo large rearrangements. This conformational change is often accompanied by a signaling event. Therefore, to regulate either the function of a protein or the activity of an enzyme, shifts in dynamic protein structure are critical. Some proteins have no need for a folding process, because they are functional in an unfolded state (figure 1.2) [35–41].

1.3 Protein microarrays

A microarray is a piece of glass or silicon on which different protein molecules or specific DNA binding sequences have been affixed at separate known locations in an ordered manner thus forming a microscopic array [42–44].

Protein microarrays are small sized devices used in biomedical applications to detect the presence and/or amounts of proteins in biological samples. Protein microarrays can be used as a multiplex method to analyze the substrates of protein

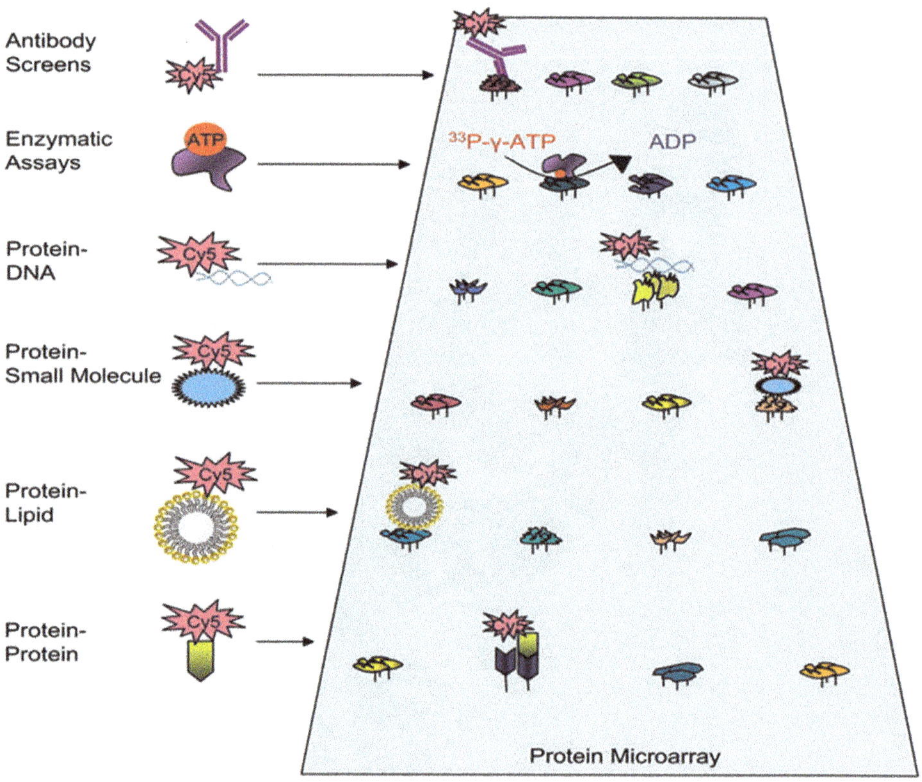

Figure 1.3. Types of molecular detection and binding pairs in protein/peptide microarrays.

kinases, to recognize the binding targets of biologically active small molecules, to understand protein–protein interactions, or to recognize transcription factor protein-activation cascades [45–47].

The peptide array was introduced as a high-throughput screening platform (figure 1.3) [48]. Current progress in the field of applications of high-density peptide microarrays is ongoing, and improved methods of manufacture upon glass slides are being examined.

References

[1] Southern E M 1998 Analysing polynucleotide sequences *US* 7811751B2

[2] Nasseri B, Soleimani N, Rabiee N, Kalbasi A, Karimi M and Hamblin M 2018 Point-of-care microfluidic devices for pathogen detection *Biosens. Bioelectron.* **117** 112–28

[3] Farjadian F, Moghoofei M, Mirkiani S, Ghasemi A, Rabiee N, Hadifar S, Beyzavi A, Karimi M and Hamblin M 2018 Bacterial components as naturally inspired nano-carriers for drug/gene delivery and immunization: Set the bugs to work? *Biotechnol. Adv.* **36** 968–85

[4] Parsa S F *et al* 2018 Early diagnosis of disease using microbead array technology: a review *Anal. Chim. Acta* **1032** 1–17

[5] Ghasemi A *et al* 2018 Optical assays based on colloidal inorganic nanoparticles *Analyst* **143** 3249–83

[6] Geysen H M, Meloen R H and Barteling S J 1984 Use of peptide synthesis to probe viral antigens for epitopes to a resolution of a single amino acid *Proc. Natl Acad. Sci. USA* **81** 3998–4002

[7] Frank R 1992 Spot-synthesis: an easy technique for the positionally addressable, parallel chemical synthesis on a membrane support *Tetrahedron* **48** 9217–32

[8] Fodor S P, Read J L, Pirrung M C, Stryer L, Lu A T and Solas D 1991 Light-directed, spatially addressable parallel chemical synthesis *Science* **251** 767

[9] Lockhart D J and Winzeler E A 2000 Genomics, gene expression and DNA arrays *Nature* **405** 827–36

[10] Vafajoo A *et al* 2018 Multiplexed microarrays based on optically encoded microbeads *Biomed. Microdevices* **20** 66

[11] Deljoo S, Rabiee N and Rabiee M 2019 Curcumin-hybrid nanoparticles in drug delivery system *Asian J. Nanosci. Mater.* **2** 66–91

[12] Karimi M, Mansouri M R, Rabiee N and Hamblin M 2018 *Advances in Nanomaterials for Drug Delivery: Polymeric, Nanocarbon, and Bio-inspired* (San Rafael, CA: Morgan & Claypool)

[13] Rabiee N and Rabiee M 2018 Biocompatibility and neuroprotective potential of encapsulated S-Allyl-L-Cysteine into PCL-based nanocarrier *Drug Deliv. Lett.* **8** 242–7

[14] Ekins R, Chu F and Biggart E 1989 Development of microspot multi-analyte ratiometric immunoassay using dual fluorescent-labelled antibodies *Anal. Chim. Acta* **227** 73–96

[15] Ekins R P 1989 Multi-analyte immunoassay *J. Pharm. Biomed. Anal.* **7** 155–68

[16] Vafajoo A, Salarian R and Rabiee N 2018 Biofunctionalized microbead arrays for early diagnosis of breast cancer *Biomed. Phys. Eng. Express* **4** 065028

[17] Kharati M, Foroutanparsa S, Rabiee M, Salarian R, Rabiee N and Rabiee G 2018 Early diagnosis of multiple sclerosis based on optical and electrochemical biosensors: comprehensive perspective *Curr. Anal. Chem.* **14** 1–11

[18] Ahmadi S, Rabiee N and Rabiee M 2019 Application of aptamer-based hybrid molecules in early diagnosis and treatment of diabetes mellitus: from the concepts towards the future *Curr. Diabetes Rev.* **15** 309–13

[19] Ferguson J A, Boles T C, Adams C P and Walt D 1996 A fiber-optic DNA biosensor microarray for the analysis of gene expression *Nat. Biotechnol.* **14** 1681–4

[20] Schena M, Shalon D, Davis R W and Brown P O 1995 Quantitative monitoring of gene expression patterns with a complementary DNA microarray *Science* **270** 467–70

[21] MacBeath G and Schreiber S L 2000 Printing proteins as microarrays for high-throughput function determination *Science* **289** 1760–3

[22] Noorirad S N, Pourghasem M, Feizi F, Abedian Z, Ghasemi M, Babazadeh Z and Rabiee N 2018 Time dependent of epigenetic effect of disulfiram on tumor suppressor gene of RASSF1A in Hela cancer cell line *Basic Res. Med. Sci.* **5** 8–13

[23] Rabiee N, Bagherzadeh M and Rabiee M 2019 A perspective to the correlation between Brain insulin resistance and Alzheimer: medicinal chemistry approach *Curr. Diabetes Rev.* **15** 255–8

[24] Hajebi S, Rabiee N, Bagherzadeh M, Ahmadi S, Rabiee M, Roghani-Mamaqani H, Tahiri M, Tayebi L and Hamblin M R 2019 Stimulus-responsive polymeric nanogels as smart drug delivery systems *Acta Biomater.* **92** 1–18

[25] Zhu H *et al* 2001 Global analysis of protein activities using proteome chips *Science* **293** 2101–5

[26] Karimi M, Mansouri M R, Rabiee N and Hamblin M 2018 Drug delivery approaches *Advances in Nanomaterials for Drug Delivery: Polymeric, Nanocarbon, and Bio-inspired* vol 1 (San Rafael, CA: Morgan & Claypool)

[27] Karimi M, Mansouri M R, Rabiee N and Hamblin M 2018 Carbon-based nanomaterials *Advances in Nanomaterials for Drug Delivery* (San Rafael, CA: Morgan & Claypool)

[28] Rabiee N, Kiani M, Bagherzadeh M, Rabiee M and Ahmadi S 2019 *Nanoparticle (NP)-Based Delivery Vehicles* (San Rafael, CA: Morgan & Claypool)

[29] Jones R B, Gordus A, Krall J A and MacBeath G 2006 A quantitative protein interaction network for the ErbB receptors using protein microarrays *Nature* **439** 168–74

[30] Rabiee N, Rabiee M, Bagherzadeh M and Hamblin M 2019 *Stimuli-Responsive Polymers: Nano-Dimension* (San Rafael, CA: Morgan & Claypool)

[31] Bahrami S, Baheiraei N, Mohseni M, Razavi M, Ghaderi A, Azizi B, Rabiee N and Karimi M 2019 Three-dimensional graphene foam as a conductive scaffold for cardiac tissue engineering *J. Biomater. Appl.* **34** 74–85

[32] Rabiee M, Rabiee N, Salarian R and Rabiee G 2019 *Introduction to Nanomaterials in Medicine* (San Rafael, CA: Morgan & Claypool)

[33] Rabiee N, Kiani M, Bagherzadeh M, Rabiee M and Ahmadi S 2019 *Smart Nanostructures* (San Rafael, CA: Morgan & Claypool)

[34] Rabiee M, Rabiee N, Salarian R and Rabiee G 2019 Nanomaterials: concepts *Introduction to Nanomaterials in Medicine* (San Rafael, CA: Morgan & Claypool)

[35] Chen D S and Davis M M 2006 Molecular and functional analysis using live cell microarrays *Curr. Opin. Chem. Biol.* **10** 28–34

[36] Stoevesandt O, Taussig M J and He M 2009 Protein microarrays: high-throughput tools for proteomics *Expert Rev. Proteomics* **6** 145–57

[37] Rabiee N, Kiani M, Bagherzadeh M, Rabiee M and Ahmadi S 2019 *An Introduction to Drug/gene Delivery Systems* (San Rafael, CA: Morgan & Claypool)

[38] Rabiee N, Kiani M, Bagherzadeh M, Rabiee M and Ahmadi S 2019 Smart drug/gene delivery systems: toward the future *Nanoparticle (NP)-Based Delivery Vehicles* (San Rafael, CA: Morgan & Claypool)

[39] Maghsoudi S *et al* 2019 Recent advancements in aptamer-bioconjugates: sharpening stones for breast and prostate cancers targeting *J. Drug Deliv. Sci. Technol.* **53** 101146

[40] Rabiee N, Kiani M, Bagherzadeh M, Rabiee M and Ahmadi S 2019 Protein-based nanostructures II *Nanoparticle (NP)-Based Delivery Vehicles* (San Rafael, CA: Morgan & Claypool)

[41] Rabiee N and Rabiee M 2018 Magnetic stimuli-responsive cobalt ferrite nanoparticle as theranostic agents for targeted delivery *Curr. Nanomater.* **3** 160–7

[42] Nour S, Baheiraei N, Imani R, Rabiee N, Khodaei M, Alizadeh A and Moazzeni S M 2019 Bioactive materials: a comprehensive review on interactions with biological microenvironment based on the immune response *J. Bionic Eng.* **16** 563–81

[43] Rabiee M, Rabiee N, Salarian R and Rabiee G Enzyme-responsive nanomaterials *Introduction to Nanomaterials in Medicine* (San Rafael, CA: Morgan & Claypool)

[44] Nasseri B, Kocum I C, Seymen C M and Rabiee N 2019 Penetration depth in nanoparticles incorporated radiofrequency hyperthermia into the tissue: comprehensive study with histology and pathology observations *IET Nanobiotechnol.* **13** 634–9

[45] Karimi M, Mansouri M R, Rabiee N and Hamblin M 2018 Advances in nature-inspired nanomaterials *Advances in Nanomaterials for Drug Delivery: Polymeric, Nanocarbon, and Bio-inspired* (San Rafael, CA: Morgan & Claypool)

[46] Karimi M, Mansouri M R, Rabiee N and Hamblin M 2018 Aptamers and pathogen-based carriers *Advances in Nanomaterials for Drug Delivery: Polymeric, Nanocarbon, and Bio-inspired* (San Rafael, CA: Morgan & Claypool)

[47] Rabiee M, Rabiee N, Salarian R and Rabiee G 2019 Bioactive nanomaterials *Introduction to Nanomaterials in Medicine* (San Rafael, CA: Morgan & Claypool)

[48] Hall D A, Ptacek J and Snyder M 2007 Protein microarray technology *Mech. Ageing Dev.* **128** 161–7

Chapter 2

Protein and peptide-based microarrays

Navid Rabiee, Sakineh Hajebi, Sepideh Ahmadi, Mohammad Rabiee, Mojtaba Bagherzadeh, Saeid Maghsoudi and Michael R Hamblin

2.1 Peptide-based microarrays

The human genome project was undoubtedly revolutionary for accelerating the discovery of thousands of new genes that were relevant to the diagnosis and treatment of critical diseases [1]. In order to untangle the function of newly discovered genes, the use of proteomics is a powerful tool. Proteomics provides a dynamic description of gene regulation, and has had a noteworthy influence on our understanding of the function of cells and the molecular pathways involved. Therefore, for a comprehensive understanding of disease pathogenesis and cellular processes, an investigation of the complete array of proteins expressed within a cell, tissue, or an organism at any particular time point is required [2–4].

A variety of proteomic techniques, including the traditional foundation method, two-dimensional gel electrophoresis (2D-GE) have been developed. The 2D-GE method was introduced as an important separation technique for protein analysis, given its potential to separate thousands of proteins in a specific tissue or cell, and to allow detection of post-translational modifications. It has been used for the universal investigation of protein expression profiles in all living organisms. However, the 2D-GE technique has a variety of limitations, including limited dynamic range, low sensitivity, not high throughput, poor reproducibility, and also has problems in analyzing very basic or acidic proteins, as well as hydrophobic and small proteins [5–7]. Therefore, 2D-GE does not provide enough information on all the different processes occurring inside the cells. 2D-GE is also time-consuming, non-automatable, and cannot provide the relative abundance of all the proteins in a proteome.

Microarray technology rapidly attracted the attention of many researchers, who appreciated its ability to reproducibly analyze protein expression and enzyme activities with good reliability in a high-throughput screening format. This advance has become increasingly essential in the postgenomic period, where the focus of investigation is being diverted from protein detection to protein functional analysis,

doi:10.1088/978-0-7503-3667-3ch2

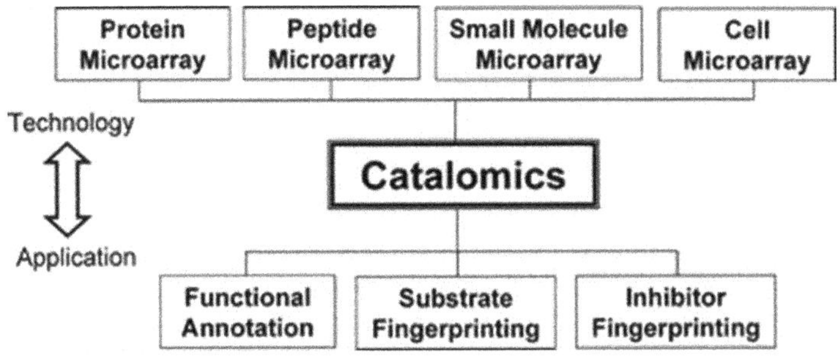

Figure 2.1. Overview of catalomics [13].

and from the study of single proteins to the measurement of many proteins and protein variants. Microarrays only require a relatively small amount of expensive reagents for most biological assays, which is another advantage [8–11].

Protein microarrays, small molecule microarrays, cell arrays, and peptide microarrays are various types of platforms used as protein-profiling arrays [12].

The term 'catalomics' was first presented in Seoul, at the Eleventh Asian Chemical Congress [13]. The chemical proteomic approach is carried out for high-throughput functional analysis of enzymes (figure 2.1).

The physical, chemical, and structural differences between various proteins, introduces a problem with their non-specific absorption on solid surfaces. This can be a limitation that prevents their reliable use in a microarray. On the other hand, using peptide substrates of enzymes immobilized in a peptide microarray could allow high throughput enzyme profiling. Currently, this 'sister platform' is being used for the high-throughput evaluation of complex protein functions. The immune response of higher organisms, and especially the analysis of immunological phenomena have been carried out using synthetic peptides for decades [14–16].

Peptides can have an almost infinite variety of structures that can be prepared using combinatorial chemistry, and can take part in molecular biology as bioactive molecules [17, 18]. Peptides do not need possibly problematic folding into stable active conformations, unlike proteins. Moreover, peptides are usually more stable and resistant to degradation compared to proteins. Accordingly, peptides can be perfect probe molecules, well established and routinely employed in a robust screening assay. Chemical synthesis has no restrictions about incorporating abnormal or unusual building blocks such as branched, D-aminoacids, organic residues, and cyclized groupings. Peptides have been employed as a technological foundation for many high-throughput assays in microarray formats, and 'catalomics' has been rapidly adopted in the last decade. Peptide arrays have also been used for rapid pharmaceutical drug screening, as well as for choosing efficient substrates, and to design inhibitors or ligands for various antibodies and enzymes [19–21].

2.1.1 Origin and development of peptide arrays

The 'one-bead–one-compound' (OBOC) combinatorial peptide library approach was first reported by Lam *et al* [22, 23]. Because a single peptide is displayed on each bead, this may be considered to be an unattractive chemical microarray, but the spatial separation allows precise analysis of binding data. The enzyme-linked calorimetric technique was used to investigate this array. Afterwards, using peptide microsequencing, beads that contained the desired isolated peptides were identified. At the same time, Houghten *et al* synthesized peptide mixtures using the peptide-based microarray method, and the local scanning method to analyze these libraries by an iterative approach [24]. From that time onwards, traditional combinatorial chemistry approaches have tended to disappear.

Schreiber *et al* took advantage of Brown's spotting method, to report the successful production of high density microarrays using non-nucleic acid biomolecules mainly containing peptides [25, 26]. This method allowed concurrent screening of tens of thousands of peptide substrates using a tiny glass surface area for immobilization. Falsey *et al* later used a peptide array that attached the N-terminal residues on glass slides with regiospecific chemical reactions [27]. Houseman *et al* prepared a glass surface covered with alkane dithiols that allowed self-assembled monolayers. Modified peptides could be immobilized and used in the array for the recognition of peptide phosphorylation by antibodies [28, 29].

2.1.2 Applications of peptide-based microarrays

In recent years, peptide-based microarrays have been introduced as a powerful tool for protein detection and functional characterization. The initial investigations using peptide microparticles were mainly concentrated on peptide–protein interactions and high-throughput screening of peptide–peptide interactions. Recently, peptide microarrays have been used for identifying and measuring enzymatic activity for the *in vitro* characterization of proteases and kinases, and for the design of enzyme inhibitors. Information about the substrate specificity and catalytic activity of a new enzyme are useful, and can shed light on the biological function of the enzyme. So far, high-throughput screening with peptide arrays has been used to profile the substrate specificity of various enzymes. Because it allows screening of thousands of enzyme–substrate reactions concurrently, peptide-based microarrays have been extensively used for substrate specificity studies, frequently for protease and kinase enzymes [30–34].

2.1.3 Kinase detection

So far, the highest number of peptide arrays that are commercially produced have been used to investigate kinase enzymes, because phosphorylation of proteins by kinases is widely used in cell biology to regulate cellular functions [35, 36]. Kinases are the second-largest group of drug target molecules after proteases, and their role is to regulate the activity of proteins by the phosphorylation of key amino acid residues that produces changes in the protein conformation. The classification of kinases is based on the type of amino acid residues that they phosphorylate, namely tyrosine, serine and threonine. An important property of kinases is their ability to detect short amino acids sequences that are contiguous to the location of the phosphorylation. Traditional detection approaches should be able to rapidly recognize the 'hit' of a specific peptide sequence. These approaches could be affinity-column selection of peptide libraries [35], phage-display peptide libraries [37], and one-bead–one-compound peptide libraries [22]. However, the SPOT technology can improve upon this by producing peptide 'macroarrays' in order to profile kinase substrates [38]. Many of the points required for high-throughput screening assays are not applicable to these techniques. On the other hand, peptide microarrays are an excellent platform, which allow concurrent recognition of several kinases in a spatially defined manner. Furthermore, a microarray-based experiment requires only a small amount of a valuable enzyme, while *in vitro*-based kinase assays usually need larger amounts of enzymes [39].

2.1.4 Recognition of proteases and other hydrolytic enzymes

One of the recent microarray applications based on peptide-based detection is the discovery of inhibitors of hydrolytic enzymes, for example, detection of protease inhibitors. Proteases cleave specific peptide bonds within proteins and other amino acid chains. Another example is the detection of inhibitors of lipase enzymes that act upon lipids and fatty acids. Protease enzymes comprise as many as 2% of encoded genes in organisms whose genome has been deciphered. Proteases are very important for cell cycle progression and for cell proliferation [40–43], which explains why their activity is highly regulated for optimum cell function. Many studies have been done on protease and other hydrolytic enzymes, as they have critical roles in cell functions. For example a range of peptide libraries have been produced, including phage-display peptide libraries [44], local-scanning libraries [45], mixture-based oriented peptide libraries [46, 47], fluorescence-quenched peptide substrate libraries [48], and labeled peptide substrate libraries [49]. Nevertheless, these approaches have not been as successful as microarray-based technologies. For this reason, a complete understanding of the variety and function of protease enzymes and substrates is not yet available. Recently, a new microarray-based strategy was reported that can detect various enzymatic activities in a microarray format utilizing mechanism-based suicide probes [50].

2.1.5 Other applications of peptide arrays

Gaining information about peptide and protein functions, their related enzymes, their interactions with each other, and the roles that they play in biological processes at a molecular level will help in the development or discovery of new drugs. For example, precise recognition of particular peptide-related sequences within a protein will help us to design enzyme inhibitors, antibody antigen pairs, enzyme substrate pairs, and to detect protein–protein interactions. Zang *et al* synthesized cyclic peptide libraries by a procedure called 'split and blend synthesis' to screen for peptide-based ligands that could bind streptavidin, which is a molecule highly specific for biotin, and is purified from the bacterium *Streptomyces*. They found that some ligands bind 1000-fold more strongly than their simple analogs [51]. One-bead–one-peptide libraries is another technique that has been used for detection of ligands for proteins, such as the Lyme antibody [52], glycosylated hemoglobin [53], RNase S-Protein [54], and immunoglobulin G [55]. In 2004, Powell *et al* described a protein-ligand binding method using a special mass spectrometry instrument called MALDI-TOF which could be used for screening 100 000 ligands per day [56].

2.2 Protein-based microarrays

In this section we will focus on the driving forces which caused a revolution in the field of protein bimolecular interactions and the corresponding microarray screening tools that could detect these interactions. Several current applications of protein microarrays will be analyzed and their current issues and challenges will be discussed. The key technologies which can overcome the above constraints, and push the industry toward higher-value applications like drug discovery will be presented [57–60].

2.2.1 Potential need for protein microarrays

Proteins play key roles in both healthy and diseased states, and act as the basic working machinery of the cell. It is important to understand how proteins interact with each other, and how they adjust to different cellular processes and signaling cascades. Protein levels inside the cell are very dynamic, in contrast to nucleic acids within the genome that contain a constant number of genes. Proteins are processed continuously, in response to external and internal stimuli, and undergo an extensive range of post-translational modifications within the body. In consequence, it is difficult to determine exactly the quantities of proteins or their exact number, which currently exist within animals or humans (or other biological systems). Moreover, families of proteins are enormously variable and have significant variability in their molecular weight, physical size, structural properties, chemical attributes, affinity constants, and their abundance within the cells can vary by many orders of magnitude. In consequence, a full categorization of proteins and their interactions is enormously challenging [61–65].

Microarrays are an attractive method to detect bimolecular interactions in a high throughput (HTP) method. Application of microarrays has been very successful in gene expression [66] and genotyping research [67] using biological samples available in only very limited volumes. Established microarray platforms still demonstrate quality control (QC) issues which are more pronounced compared to DNA arrays. There will still be challenges involving protein denaturation and QC, when microarrays are used for screening proteins.

Several applications have resulted in a need for better and more reliable protein screening tools.

2.2.2 Protein therapeutics

During this current decade the discovery of antibody-based drugs is anticipated to accelerate even further. It is predicted that various products will come to clinical trials in the next several years, leading to even hundreds of monoclonal antibodies coming onto the market [68]. Nearly 20% of biological products currently sold are monoclonal antibodies [69]. Therefore, microarrays are often used to observe these high-affinity binding molecules and their involvement in drug discovery pipelines is huge. The sequencing of the human genome was another driving force, because the number of drug-like targets has increased sharply requiring a large number of molecules. Screening of these new molecules is still at a very immature stage, and before their potential can be hitched for drug applications, much work will need to be done.

2.2.3 Clinical diagnostics

All over the world infectious diseases such as hepatitis C (HCV), herpes, HIV, and novel coronavirus (COVID-19) are increasing rapidly with several million new cases yearly. Infectious epidemics can affect the entire community in both developed and developing countries. According to the World Health Organization, over 170 million people are infected with HCV [70], and other new infections are increasing at a rate of 3–4 million yearly. Many diagnostic tests are based on PCR formats or else on ELISA formats. PCR-based tests are expensive, and the cost is a factor that prevents them from being extensively used in all countries, despite the fact that they is more sensitive than ELISA. The ELISA technique is a standard method with sensitivity ranging from 70%–95%, and continues to be widely used. One of the most effective ways to screen infectious diseases is the use of microarray-based platforms that allow screening of antigen–antibody interactions using smaller amounts of samples. A protein microarray containing antigens is a low cost approach to search for a variety of diseases. The need for such screening platforms is bound to grow (provided costs are kept to a minimum) because of the increasing number of potentially infected individuals. In addition, screening for biomarkers for auto-immune diseases, cardiac disease and diabetes type I are drawing much attention to the area of protein microarrays.

2.2.4 National security and biological threats

There is currently a growing demand for methods to detect pathogens in the environment, under battlefield conditions, and in food, necessitating the discovery of novel technologies that can rapidly identify bacteria and other toxins [71]. The ability to assemble proteins onto solid substrates equipped with microfluidics (and similar fluid-handling systems) that can be automated is crucial. Proteins or antibodies can survive under harsh conditions with appropriate microarray surface immobilization chemistries. Moreover, novel binding approaches such as protein imitators, aptamers or synthetic molecules could be used [72].

Western countries are committed to national security, thus additional funding will likely be made available to develop new sensing instruments in the future. All these activities will have a positive effect on the development of microarray-based protein technologies.

2.2.5 Recent applications of protein microarrays

Most applications of conventional glass-based arrays have been used for disease-related proteomics [73]. Significant experience earned from glass-based DNA arrays by the research community can be applied to various other types of protein arrays [74].

Samples can consist of all types of human bodily fluids, such as serum [75], synovial fluid [76] sweat [77], urine and saliva. Also, proteomic studies in the placenta in the field of gynecology have also been reported [78]. Although, it is expected that the use of this kind of prenatal analysis will grow, ethical concerns still remain. Since most government and pharmaceutical programs have concentrated on cancer to the exclusion of other diseases, it is not surprising that more research has been devoted to various cancers.

Screening antigen–antibody and carbohydrate–protein interactions has been developed using peptide [79] and oligosaccharide [80] microarrays. The results of studies on carbohydrate arrays, although still in the early stages, are promising. Several combinations of peptide and protein microarrays for analysis of protease substrates [81], cytokine activities [82] and detection of bacteria or toxins [83] have also been successfully deployed. Recently, some new studies have been reported using cell membrane preparations [84] and GPCR (G-protein coupled receptor)-related activities [85]. It is very difficult to assemble a sufficient biologically functional membrane layer (which accurately mimics the natural system) on a solid substrate, which is an elegant way to study receptor–ligand interactions. The surfaces are easily affected by small changes in pH, temperature, humidity among other factors, and the stability of such membranes is still in question. Peptide-based platforms can be used for certain preliminary analyses, and these systems can progress to biomedical or bioelectronic applications. Currently, one of the most active fields of research is in clinical diagnostic applications using protein arrays in the discovery of new biomarkers [86]. Although numerous studies are currently being carried out in laboratories, universities, and industry, today none of these tests are yet ready for the diagnostic market. Advances in technologies resulting in the development of clinical assays are expected in the next 3–5 years.

2.3 Problems and challenges

The final goal of any screening experiment is to determine exactly the thermodynamic properties and kinetics of the biomolecular and behavior, after that integrate the information into downstream applications such as biomarker analysis or drug discovery. The logical design of medicinal drugs and tools for diagnostics requires sound knowledge of biomolecular behavior.

To define the relationship between two interacting molecules (e.g. proteins), a rate constant can be produced on any screening platform. The actual question (principally for drug discovery purposes) is: how relevant is the information and does it closely mimic the processes happening in natural biological systems and their surroundings? In diagnostic applications, it is essential that the envisioned assay kit, relying on an interaction between immobilized surface probes and environmental biomarkers or targets in biological specimens, can accurately identify the patient's condition without any false positive or negative results.

The efforts in modern drug discovery require the screening of millions of compounds in a high throughput format before any new therapeutic candidates can be identified. The same is true in the case of diagnostic tests, which may need to

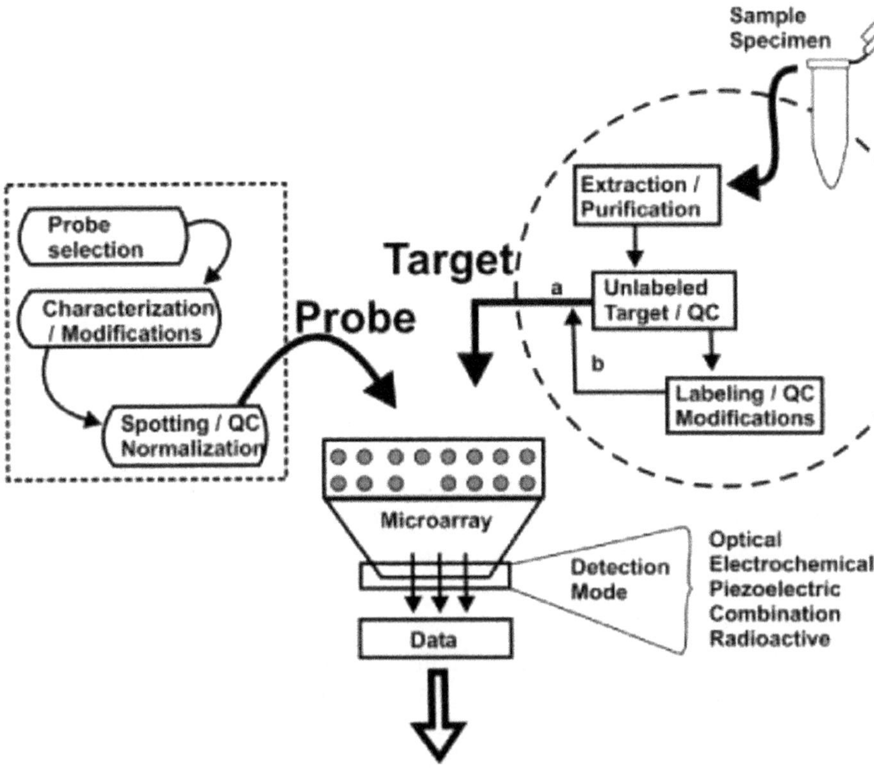

Figure 2.2. Conventional microarray processing challenges are present at all stages of the flowchart. All the processes occurring during the microarray experiment starting from sample probe and sample preparation steps are shown: preparation, interactions on microarrays within the square and the circle. Significant—detection and data analysis.

be carried out in several hundreds of thousands for certain disease types. The potential pitfalls (figure 2.2) in microarray-related screening include, although are not limited to:

- sample preparation and handling (probe and target);
- microarray platform;
- recognition technology;
- data analysis.

2.4 Sample preparation and handling (probe and target)

Before any single microarray experiment can be manufactured on a commercial scale, several challenges need to be overcome at the sample collection and processing steps. The main aims of the experiment can be defeated if the chosen probes and targets are not properly characterized. Furthermore, suitable quality control measures should be introduced, especially during any chemical modification steps (tagging or conjugation of surface-anchoring groups) on the targets and probes. Since the information analysis is based on the identified signal (which is correlated to expression levels, etc), it is important to exactly normalize each target sample and probe, therefore the results will be representative of the true bimolecular expression level. Based on the literature, most initial sample preparation steps include procedures which are laboratories. Though some binding reporter methods can work with unlabeled targets, most microarray formats require signal reinforcement. As a result, extra tagging of the targets is required. The use of secondary marked targets is crucial in special sandwich techniques (applicable to diagnostic studies where primary marking of the biomarker target is not easy in patient samples). At each stage, understanding the basic principles of operation of the purification and extraction methodology will lead to reducing the overall uncertainty. Some procedures consume quite a significant amount of time, such as multiplex experiments with many targets. In this case the use of automated systems will be vital. During the processing some probes may be easily denatured. Drying of spots can be challenging but certain types of humidity chambers can be used to minimize this problem. Moreover satisfactorily spotting the probes and targets can be complicated. How well trained the end-user is and how good is the experimental design will be the most pressing questions.

2.4.1 Microarray platforms

The price of the microarray platform is much lower than the cost of reagents (buffers, probes, etc). Amazingly, the research community still cannot fully appreciate the importance of the microarray surface chemistry (the foundation of a microarray), and the need for its efficient characterization prior to any use to conduct protein–protein interaction or any other biomolecular studies. If the microarray surface is appropriately prepared, most of the existing problems facing microarrays, such as non-specific interactions, denaturation of biomolecules, spot heterogeneity, and microarray smearing can be addressed effectively.

2.4.2 Detection technologies

The existing types of reporting read-outs involve optics, electrochemical, piezoelectric or radioactivity. Optical methods (fluorescence-based detection systems [87], surface plasmon resonance (SPR) [88], waveguides [89], etc) and piezoelectric techniques [90] have sufficient sensitivity to be used in protein microarrays. Advances in other areas including improved binding interactions, use of tags with improved optical properties, higher signal-to-noise ratios, and microarray surfaces with low background, will help to enhance the sensitivity of these techniques. A combination of the above techniques combining measurement of the viscoelastic, optical and electronic properties of the biomolecular interactions can provide additional information on the packing compression, orientation of the proteins and also kinetic information.

2.4.3 Data analysis

It is difficult to control and monitor binding interactions in several hundreds or thousands of different specific locations. In addition, the comparisons required for conducting the statistical analysis result in the generation of many additional datasets. Despite the introduction of advanced data analysis software [91], the true biological importance of the data is still difficult to ascertain. Moreover, an improper orientation of the microarray or the presence of dust particles could throw the information analysis software program off balance. In addition, the information analysis is also at the mercy of all the problems that arise previous to data collection. Additionally, analysis regression software, mostly used for kinetic studies, can provide information on k_{on}/k_{off} rates. For end-point trials, where real-time kinetics is not required, other challenges are presently being addressed.

2.5 Potential solutions: enabling technologies and advancements

There are currently a number of interesting new technologies that can be used to overcome the drawbacks of microarrays (both protein and DNA). Self-assembly methods to improve the biocompatibility of protein microarrays on gold and metal oxide surfaces are useful. Schaeferling's research group focused on these technologies and reviewed most of them. This provides a comprehensive summary of all existing surface chemistry options that can be applied to semiconductors, glass and plastic substrates for efficient binding of proteins and other biomolecules. Prolinx developed small molecule chemical affinity systems which can be used to expand protein arrays in both two- and three-dimensional formats [74, 92, 93].

There have been recent reports describing the observation of real-time biomolecular interactions on gold-based substrates by the research groups headed by Professors Corn and Knoll [94, 95]. The plasmonic properties of gold surfaces allow some interesting new optical technologies such as surface plasmon resonance (SPR), surface plasmon microscopy (SPM) [96] and surface plasmon-enhanced fluorescence spectroscopy (SPFS) [97]. These technologies can be used with either tagged or label-free detection modes, and are therefore useful when dealing with proteins in which the addition of a label may significantly alter the properties of the biomolecular interaction. The SPFS technique has been to develop a real-time diagnostic instrument for the measurement of blood analytes. So this technology is already allowing biomolecular screening in the realm of clinical diagnostics. Fluorescence-based detection systems still dominate the microarray field and in future it can be predicted that LED-based scanners will considerably lower the cost of assays. The use of novel nanoparticle-based systems such as quantum dots [98] and the principles of resonance light scattering [99], can also be used to increase the signal, but the cost of these materials could still be an issue.

2.5.1 Protein microarray surface chemistry and coupling schemes

An important factor in determining the general performance of a microarray or any high throughput (HTTP) screening tool is efficient surface immobilization of probes. Improperly oriented immobilized probes or denatured probes can dramatically impact the efficacy of downstream biomolecular binding events. In this section the coupling schemes for immobilizing biomolecules on different substances are described, including gold, glass, polymer and semi-conductor surfaces. The substrates and the different types of surface chemistry should be tailored, and various schemes for manipulating the probes using labels will be mentioned.

Different approaches for fabricating protein microarrays on an extensive range of substrates will discussed in this review. The exact way that proteins are immobilized on solid substrates will govern the functional properties of a protein microarray. Obtaining high-quality data on biomolecular interactions, which can be used to predict how proteins behave in their biological environments, is the overall goal. Currently, protein interactions are performed in a wide range of formats and surfaces. Two approaches are proposed for expanding and improving protein microarrays. For the first approach, the properties of the microarray surface are designed to immobilize unmodified (pristine) biomolecules, while in the second approach, the biomolecules are modified with labels to effectively anchor them to available sensor substrates in the microarray. The interpretation of information remains a fundamental challenge, regardless of which approach is chosen for a specific application. Because proteins have such a wide variety of structural, chemical, and biological properties, approaches to protein immobilization that work for one type of protein cannot necessarily be extrapolated from one system to another. Carrying out normalization studies with a single protein array system (corresponding to activity and probe surface concentration) has to be repeated with an extensive range of probes with different functional properties. Currently, quality control issues still plague sample preparation and microarray designs. Troubleshooting these processes to achieve the maximum overall benefits of such microarrays in FHTS is enormously time consuming.

A 'one-approach-fits-all' approach is highly improbable for screening all the different types of proteins. Current practical approaches can be used for an extensive number of protein families without compromising their overall ability to bind to surfaces. We will discuss in detail how protein chips can be manufactured using diverse immobilization chemistries on gold, glass, polymer and other specific materials (semiconductor surfaces and lab-on-a-chip). Various strategies will be described for the use of effective tags for increasing the degree of probe immobilization and preventing protein denaturation.

2.5.2 Background and recent state of biomolecule libraries

Large pharmaceutical companies, biotechnology companies, and academic research laboratories currently have available libraries of both pristine proteins and proteins conjugated with different labels (biotin, Avi-Tag, Strep-Tag, His-Tag, etc). The surface immobilization strategy to attach proteins to microarrays might be chosen on the basis of the existing chemical state of the libraries, because laboratories will probably be unwilling to alter their molecules to accommodate the surface immobilization requests of different microarray platforms. Biomolecule libraries exist in all of the microarray formats that will be described in the present book. Many varieties of proteins have sensitive functional groups in their side chains which can be used for attaching these molecules onto microarrays as a starting point. For biomolecules conjugated with labels, when compared to pristine probes, immobilization approaches appear to have a higher surface coupling efficiency. The problem in this case is that the interactions between the biomolecules and the conjugated probes may alter, thus affecting the thermodynamic and kinetic properties of the binding. Currently, numerous surface chemistry methods that function regardless of the nature of the probes are available. The goal is to target reactive sites which do not denature the molecules or affect the affinity of the complementarity-determining regions (CDRs) of the probes after immobilization on the surface.

2.5.3 Microarrays based on glass substrates

Most of the existing studies on protein interactions on microarrays have been carried out according to formats and surfaces that have been previously utilized for DNA arrays. Glass slides have been used to immobilize proteins, antibodies and antigens. The glass surface is modified by epoxy- or aldehyde-terminated alkylsilanes or by polylysine. The final activated glass surface (microscope slide format) can be spotted at a chosen density with biomolecules. The recognition mode for such systems is essentially based on optical methods such as fluorescence.

2.5.4 Surface modification

The responsive coatings on the glass surface are mainly prepared according to self-assembly procedures. The hydroxylated surfaces act as substrates for the construction of self-assembled monolayers (SAMs) using alkylchlorosilanes or alkylalkoxysilanes. The *in situ* formation of polysiloxane is the driving force for this self-assembly reaction, which is attached to the surface silanol groups (SiOH) via Si–O–Si bonds. These thin organic films alter the biocompatibility of the surface and protect the proteins from denaturation and from structural changes occurring during the immobilization step.

An enormous variety of silanization reaction conditions have been described in the literature, varying in reaction time and temperature, which are governed by the stability of the terminal functional groups [100]. For instance, under mild conditions (room temperature for 30 min) monolayers of aminopropylsilanes can be formed [101]. In figure 2.3, some silane-reagents commonly-used for the derivatization of glass surfaces are shown [102]. Glass, quartz, silicon oxide, aluminium oxide, mica, zinc selenide and germanium oxide have all been successfully silanized to form organosilicon monolayers [103]. Accordingly, silicon substrates covered with silane monolayers, led to a new method for modification of the electronic properties of semiconductor chips [104] and for preparing new interfaces in the field of bioelectronics.

2.5.5 Current state of glass-based protein microarrays

There are some challenges and issues in relation to spot morphology, including homogeneity of signal throughout the spot, microarray smearing, and non-particular adsorption, which may not have been completely taken into consideration in spot protein microarrays. In addition, it is hard to compare the experimental results gained under the same conditions in various laboratories using a similar set of

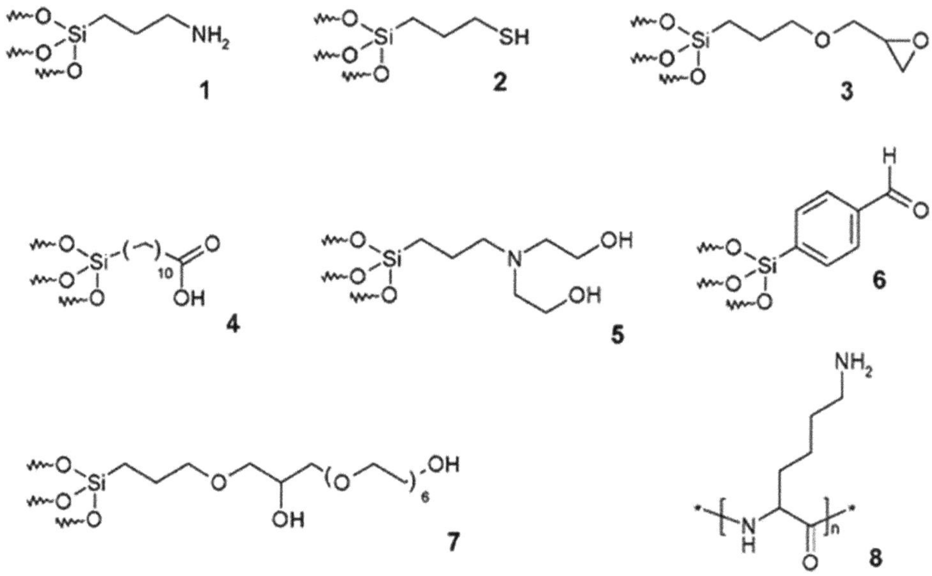

Figure 2.3. Some reagents for derivatization of glass surfaces [102]. Copyright 2002 John Wiley & Sons Ltd.

arrays. Reproducibility of data, which in its current state can cause extreme difficulty for accurately validating the biomolecular interactions, is considered a highly important issue. Accurately validated biomolecular interactions are the foundation for many drug discovery research studies. Initial studies were conducted using polylysine arrays to observe the interactions of 115 antigen/antibody pairs using complex protein mixture solutions [105]. Despite obtaining successful interaction data, in some studies between 20% and 50% of microarrays showed smearing and high background still remains an issue.

Microarray smearing is a common artifact that occurs in most glass-based arrays, and can cause cross-talk between neighboring probes. Moreover, electrostatic charges present on the glass surface can denature proteins, and in addition it is difficult to estimate the exact level of electrostatic charge. This can create false results. In another investigation, large-scale protein–protein interactions were successfully investigated using aldehyde-coated glass slides [25]. These slides could discriminate between specific and non-specific interactions, but 'quenching' or the need to neutralize the residual reactive groups on the slides remains a problem. Other issues include problems that can be posed while using small molecules to block the free surface of a chip. The use of bovine serum albumin (BSA) in order to minimize non-specific interactions is better with bulky BSA molecules. Critical problems are caused when probing many thousands of protein samples spotted onto a single microarray slide with spot morphology, while on a laboratory scale it is not considered an important issue. Fluorescence signals may originate from the spots themselves or from the adjacent zones (in the case of inhomogeneous spots) potentially affecting the data. Furthermore, non-homogeneous surface chemistry inside the spots can lead to differences in fluorescence signal inside that same spot.

Highly sophisticated data analysis is required to prevent poor averaging factors within the analysis step, which could cause aberrant protein expression data for the protein of interest.

Joos *et al* [106] described the direct discovery of auto-antibodies in human sera by taking advantage of a single immunoassay experiment that involved spotting 18 different antigens onto aldehyde-coated slides. In order to prepare a high-throughput microarray-based enzyme-linked immunosorbent assay (ELISA) [107] activation of a 96 well glass plate with Teflon coated wells (Erie Scientific) using aminopropylsilane was conducted. Protein antigens were then applied to predetermined spots. The authors also considered arrays with 144 wells each using a 36 capillary based print head. This array arrangement could be a compromise between conventional micro-titer plates and high-density microarrays. The antibodies could be arrayed on the aminosilated glass surface using a micro-molded hydrogel stamper [108]. Some approaches have used PDMS stamps in micro-contact printing [109], or photoreactions in a mixture with masks [110] for the patterning of proteins onto glass surfaces.

A glass microarray coated with aldehyde or poly-lysine containing 146 distinct antibodies was used for the profiling of cancer cells using a cellular lysate. The target proteins were tagged with Cy3 or Cy5 dyes and no reinforcement techniques such as ELISA were needed for monitoring the interactions [111]. For instance in one interesting study, Schweitzer *et al* studied cytokine secretion from human dendritic cells induced by lipopolysaccharide or tumor necrosis factor-a. They used microarrays containing 75 antibodies against cytokines attached to chemically-derivatized glass slides [112]. Snyder reported the advantages of nickel complex-coated surfaces over conventional aldehyde glass slides for a proteome chip containing 6566 GST yeast proteins spotted onto a glass slide. In this specific case, the proteins were attached through their His-Tags and it was assumed that the binding sites were oriented at a distance from the surface [113]. This example suggests that the use of labels can help to control the immobilization of the biomolecules on the surface.

2.5.6 Microarrays based on gold substrates

Gold-based substrates have been used to monitor protein interactions over a rather long history. Surface plasmon resonance techniques are most effective with gold films or nanoparticles, as compared to other metals. Thin gold films 10–200 nm can easily be deposited onto solid supports similar to glass or silicon wafers by evaporation or sputtering in an ultrahigh vacuum (UHV). Polycrystalline gold surfaces have also been used [114, 115]. Meanwhile the adhesion of silicon or gold onto glass is rather poor, and pre-coating of a thin layer of Cr or Ti (1–10 nm) is usually essential prior to gold deposition.

2.5.7 Surface modifications

Hydroxy-terminated alkanethiols with oligoethylene glycol units [116, 117] and mannitol groups [118] can be employed to lower non-specific interactions with enzymes or proteins. This makes them good candidates to create hybrid monolayer-based protein array platforms. For further information about surface modifications with the capability to resist protein adsorption, please refer to the extensive review by Whitesides [119].

2.5.8 Current state of gold-based protein microarrays

Gold-based protein microarrays have been used most extensively based on biotin–streptavidin binding. An important reason for the extensive use of biotin–streptavidin technology, is that it is possible to readily biotinylate many different biomolecules. However, regarding the use of biotin-conjugated molecules, the following two conditions are essential: (a) the properties of the biotin moiety are not affected by the attached biomolecule; and conversely (b) the physical and biochemical properties of the biomolecules are not affected by the biotinylation step.

The relative ease of attaching biotin to organic molecules with thiol groups bolsters the use of gold substrates as a platform for constructing protein arrays. The streptavidin surface can be used for immobilizing a widespread range of biotinylated molecules (proteins, polysaccharides, nucleic acids, etc), which are attached on top of the underlying biotin-terminated SAM.

Single-component SAMs comprising biotin attached at one end are used, and due to steric limitation caused by the tight biotin layer, avoid the formation of streptavidin monolayers. The closely-packed biotin coating protects individual biotin units and hinders the efficient binding of large streptavidin molecules. This problem can be overcome by using any spacer molecules or by using surface binding schemes that may have lower efficiency.

Different investigations on the interaction of biotin and streptavidin monolayers in various sensor designs have been made by the groups of Knoll [120] and others [121]. Knoll and colleagues have shown that by using a 1:9 molar ratio of hydroxyl-terminated thiols to biotin they could generate an optimum monolayer of streptavidin. Various studies on the composition of biotinylated thiols and diverse spacer molecules showed they had two functions. Firstly, they can act as effective spacers between the biotinylated components. Secondly, they minimize the non-specific interaction of the sensor substrate with streptavidin. Other study groups [122] have prepared an optimum streptavidin monolayer using various streptavidin protein mutants that can accommodate 10%–25% biotin on the surface.

The reaction at the surface between a hydrophobic interface including biotin binding units linked to amino groups and carboxy groups is another approach to produce a well-packed surface of streptavidin [123, 124]. A densely packed streptavidin surface, where the superficial biotin density is between 25% and 30% can be obtained. Also, the amount of mixed SAM layers is within the same range. In the application of this surface in screening patients affected with Lyme disease and Borreliosis, there were no blocking solutions used in this study, and highly reproducible results with a high signal-to-noise ratio were obtained, because the array surface was highly homogeneous (i.e. had a uniform probe packing density) [125].

Although the SAM compounding method is comparatively easy to perform, and can facilitate the manufacturing process, the above-mentioned techniques can be used to assemble a streptavidin surface. The problem that exists here is that the molecules are not readily available and have to be synthesized from scratch. The synthesis reactions of these molecules usually have a poor yield and are non-trivial to carry out. This problem can be remedied using existing carboxy-terminated thiols to manufacture the streptavidin sensor surface. The SAM compounding method has been much used for the screening of patient sera, antibodies and other difficult proteins due to the good biocompatibility of the SAM interface.

Numerous reports have described the design of proteins immobilized on gold-covered surfaces using the chemical selectivity of mixed monolayers [126], lithography [127], photoimmobilization [128], and microcontact printing [129]. The direct imaging of protein molecules attached to gold surfaces can be carried out using scanning tunneling microscopy (STM) [116] or atomic force microscopy (AFM) [129]. In the second case, the thiolated protein was directly absorbed on the surface allowing direct communication between the proteins and the metal surface, however, some protein denaturation was observed in the study.

Gold-coated electrodes can also be used as amperometric biosensors in the production of enzymatic electrodes, where enzymes are immobilized on conducting polymers such as polypyrrole or polythiophenes [130]. The biosensors can also be immobilized via functionalized SAMs [118], or by direct absorption onto the electrode surface [131]. Two noticeable examples consist of a glucose sensor employing electrochemical recognition of enzymatically-produced hydrogen peroxide (glucose oxidase) [121], or the screening of neurotoxins with immobilized acetylcholinesterase in a competitive assay (using acetyl-thiocholine) [132]. Nevertheless, neither of these methods is likely to be used in a microarray format for HTP studies in the near future.

2.5.9 Microarrays on polymer substrates

The previous discussion has concentrated chiefly on two-dimensional B–D surface constructions. Some sensitive proteins may not be able to stay intact in the harsh conditions of a 2D surface, because the structure and conformation of a protein are highly dependent on the environment. As a consequence, other designs have been proposed based on C–D three-dimensional surfaces with a fluidic surrounding that closely mimics the biological environment of sensitive proteins. Hydrogels and polymer membranes are some prominent examples of 3D-based systems. The underlying principle of the 3D system is that the conformation and the activity of the protein are preserved by the fluidic environment. The higher probe density and greater surface area are other benefits of the 3D architecture. The first substrates used for the preparation of microarrays were polymer membranes such as nitrocellulose [125, 133] or aminated nylon [123]. For more than two decades in microbiology laboratories, these platforms have been extensively employed as blotting membranes, using radioimaging or fluorescence to detect samples on these supports.

Most of the studies using hydrogels have employed downstream detection methods such as mass spectrometry in proteomic investigations. The special benefits of these platforms are simple sample preparation and low overall cost, because antibodies and proteins are directly deposited on the blotting membranes after being separated by electrophoresis. Extensive application of SPR technology by Biacore in hydrogel arrays on gold substrates has led the field of real-time protein interaction studies in the last few years. Polymer hydrogels and polymer brushes on glass substrates with a traditional microarray arrangement have also been reported, however, their use is still limited.

2.5.10 Surface modifications

Functionalized polymeric membranes like cellulose or nylon can undergo immobilization of biomolecules caused by electrostatic absorption, and this also occurs for hydrogels based on polyacrylamide platforms. Polymer-based microarray studies have faced some difficulty in achieving trustworthy and high-quality data. During the past decade however, self-assembled monolayers modified with dextran have become systems of choice for studying biomolecular interactions. The extensive use of this 3D construction was mostly a result of the achievements of the commercial SPR instrument manufacturer, Biacore. Researchers at Biacore used recognized self-assembly protocols onto gold to synthesize these dextran surfaces [134]. The sensor surface was modified with epoxy-terminated thiol monolayers to allow the coupling of dextran polymers onto the surface, by reaction with carboxy-methyl-modified dextran. This was seen as a versatile reaction platform for covalent attachment and activation of molecules and immobilization of ligands. The resulting layer consisted of a 3D hydrogel and a 2D monolayer on the gold surface, which was non-crosslinked, highly flexible and could expand in a physiological buffer from 100–200 nm.

Despite the fact that 3D surfaces have many advantages, there are still the problems of non-specific interactions, high background signals, and difficulties with mass transport. The 3D dextran surfaces can alter the inherent kinetics of the interaction by restricting mass transfer. Diffusion effects have an important role in the establishment of local concentration gradients within these 3D structures. The result of incorrect calculations obtained during using surface plasmon resonance measurements for real-time kinetic studies may lead to such local gradients. In the following sections we will discuss protein immobilization with the assistance of photochemically-activated polymers or by functionalized polymer brushes.

2.5.11 Current state of polymer-based microarrays

The dextran surfaces act as a base, and can be modified with amine groups, streptavidin, thiol groups, or aldehyde groups in order to attach various proteins to the chip surface. Success has been achieved in the commercialization of numerous sensor chips for therapeutic and diagnostic applications, in fields such as HIV. Analysis is often based on numerous monoclonal antibody–antigen interactions [135].

Sensors can be useful in both acidic and basic environments, and since these sensor chips are fairly robust, they can be used multiple times without any substantial loss of the surface density or activity of the ligands. The combination of self-assembled monolayer techniques in a 3D microenvironment is another new method for the synthesis of functionalized dendrimers on solid supports such as glass. The dendritic structures can be activated by NHS esters or isothiocyanate groups on the surface [136]. Newly developed technologies for functionalization have used photografting, or the (co)polymerization of vinyl alcohol, styrene or methacrylate monomers onto the surface of a glass chip [137]. The optical properties of the chip surface can be designed to reflect light or to have sophisticated scattering effects. The refractive index of the surface can be adjusted by addition of low refractive layers (polymers, SiO_2) or high-refractive index materials (TiO_2) [138]. In order to allow the use of highly sensitive luminescent-based assay systems, this method has been adapted to prepare extremely reflective microarray surfaces at chosen wavelengths (excitation and emission).

A comparative study involved the screening of 11 various array surfaces on polymer gel and silane-coated glass slides was performed using five different monoclonal antibodies in a 20×18 matrix. In this work, all the important characteristics of the procedure were studied, including handling, ease of chip preparation and storage, reproducibility, detection limits, and signal intensities. They used reflective slides with an Amersham Biosciences, mirror-like surface C-aminopropyl-triethoxysilane, to demonstrate reproducibility and a high level of uniformity, while slides covered by polyacrylamide with high sensitivity were also studied. The authors concluded that the choice of the array material governed the properties of the probe biomolecules and the specific requirements of the device [139, 140].

Other studies have described the use of a multiplexed sandwich assay on nylon membranes, or custom-made antibody arrays on nitrocellulose for the determination of cytokines in human serum or other media [141].

2.6 Specific formats: microfluidic devices and integrated semiconductor chips

Over the last decades, lab-on-chip technology employing microfluidics, has very much advanced, and is now used in the field of proteomic research for parallel analysis and detection of several samples in an automated fashion. However, conventional microarrays are simpler than these devices, and whereas they are useful for small-scale sample sizes, nevertheless their throughput is rather low. The user friendliness of these microfluidic systems is slowly improving, considering that these systems are being transferred from university labs to commercial technology [142]. Despite these limitations, the requirement for only a very slight amount of urine or human sera to perform rapid diagnostic testing is an important benefit of this method. This technology will be a groundbreaking revolution during the present decade (2020s) to enable nucleic acid testing. Caliper (http://www.calipertech.com) and Nanogen (http://www.nanogen.com) are working to incorporate these technologies in lab-on-a-chip SNP genotyping systems. Although most microfluidic formats have focused primarily on DNA interactions, if the surface immobilization strategies are changed, such a system could also be used to detect antibody–antigen interactions or protein–protein.

Many microfluidic systems use chips with embedded microchannel systems prepared using classical etching techniques. Processing technologies use systems based on glass or silicon substrates, although Nanogen incorporates silicon wafer biochip units inside a plastic cartridge. The biomolecular interactions using highly sensitive fluorescence identification approaches are observed by planar waveguide glass chips from Zeptosens, uniting microfluidics combined with high refractive index material (Ta_2O_5) implanted onto the glass support [89]. Also, the Flow-thru-Chip™ from MetriGenix is based on a three-dimensional microchannel glass chip [143]. CombiMatrix [144] is a proprietary porous 3D layer connected to the active side of a semiconductor surface for the immobilization of biomolecules. Currently at Zyomyx, another microfluidic protein biochip platform based on silicon is under development [145].

Polymeric materials such as poly-(methyl methacrylate) (PMMA) have been used to create microfluidic microchannels, because of the ability to be chemically derivatized with reactive groups such as methyl ester groups on the surface of the PMMA in order to immobilize biomolecules [146]. The combination of microfluidics with ionization mass spectroscopy (ESI-MS) [147] or with micro-capillary electrophoresis [148] are two examples of potential benefits of miniaturization, which can be combined with other detection technologies for efficient analysis of protein interactions.

The semiconductor industry technology is used in the manufacture of micro-electrode structures on silicon wafers in the field of seamless microarrays. Attaching biomolecules to functionalized monolayers or to conductive polymers can integrate them with silver, gold or platinum electrodes. Therefore, a voltage could be useful for the selective immobilization of the physiologically charged biomolecules on such electrodes. One example of such an approach is the microarray system of Nanogen, although a conventional optical method is still used for the chip read-out [149]. There are other advanced methods that are currently only used for electrochemical recognition of DNA hybridization in combination with redox-active tags [150] (such as Motorola's eSensor™ DNA detection system) or Intercalators [151]. At the moment, these technologies have focused mainly on DNA arrays, however, in the future, these techniques could be used for detecting antibody–antigen or protein–protein interactions. These technologies could possibly be the key to other applications (e.g. BIONCHIP) such as capacitance measurement or impedance spectroscopy on a surface [152]. The electrochemical detection methods are similar to the luminescence-based assays, and can also be incorporated into ELISA methods. By coupling enzymes such as alkaline phosphatase or horseradish peroxidase into the chips, target molecules can produce reactive products that can be detected by impedance spectroscopy or by amperometric approaches [153, 154]. Recently, a wholly incorporated electronic CMOS (complementary metal oxide silicon) biochip based on enzyme-linked sensor technology has been introduced by Infineon [155].

2.7 Site-specific immobilization of antibodies

It is a well-known property of antibodies to be able to bind reversibly to protein A-coated surfaces. For high-throughput immunoassay applications, antibodies of the

IgG type can be connected to protein A surfaces in order to take advantage of their Fc-domains leaving the heavy chains for antigen binding. Under these circumstances, the Fab-domain (antigen binding region) must be oriented above the surface in order to be readily accessible for the antigens. Protein G can be used for antibody immobilization in case some monoclonal IgG antibodies do not bind to the protein A surface. A recent review by MacBeath covered various immunoassay strategies for detecting protein interactions [156].

Another covalent-linking method is the use of hydrazide-terminated monolayers that bind only to aldehyde groups that are only found in the glycosylated segment of the heavy chains of antibodies, but are not present in the protein chains of the Fab-domains.

References

[1] Venter J C, Adams M D, Myers E W, Li P W, Mural R J and Sutton G G *et al* 2001 The sequence of the human genome *Science* **291** 1304–51

[2] Nasseri B, Soleimani N, Rabiee N, Kalbasi A, Karimi M and Hamblin M 2018 Point-of-care microfluidic devices for pathogen detection *Biosens. Bioelectron.* **117** 112–28

[3] Farjadian F, Moghoofei M, Mirkiani S, Ghasemi A, Rabiee N, Hadifar S, Beyzavi A, Karimi M and Hamblin M R 2018 Bacterial components as naturally inspired nano-carriers for drug/gene delivery and immunization: Set the bugs to work? *Biotechnol. Adv.* **36** 968–85

[4] Parsa S F *et al* 2018 Early diagnosis of disease using microbead array technology: a review *Anal. Chim. Acta* **1032** 1–17

[5] Rabilloud T 2002 Two-dimensional gel electrophoresis in proteomics: old, old fashioned, but it still climbs up the mountains *Proteomics* **2** 3–10

[6] Gauss C, Kalkum M, Lowe M, Lehrach H and Klose J 1999 Analysis of the mouse proteome. (I) Brain proteins: separation by two-dimensional electrophoresis and identification by mass spectrometry and genetic variation *Electrophoresis* **20** 575–600

[7] Ghasemi A *et al* 2018 Optical assays based on colloidal inorganic nanoparticles *Analyst* **143** 3249–83

[8] Vafajoo A *et al* 2018 Multiplexed microarrays based on optically encoded microbeads *Biomed. Microdevices* **20** 66

[9] Karimi M, Mansouri M R, Rabiee N and Hamblin M 2018 *Advances in Nanomaterials for Drug Delivery: Polymeric, nanocarbon, and bio-inspired* (San Rafael, CA: Morgan & Claypool)

[10] Rabiee N and Rabiee M 2018 Biocompatibility and neuroprotective potential of encapsulated S-Allyl-L-Cysteine into PCL-based nanocarrier *Drug Deliv. Lett.* **8** 242–7

[11] Vafajoo A, Salarian R and Rabiee N 2018 Biofunctionalized microbead arrays for early diagnosis of breast cancer *Biomed. Phys. Eng. Express* **4** 065028

[12] Yi H, Mahesh U and Shao Q Y 2006 Microarray: a versatile platform for high-throughput functional proteomics *Comb. Chem. High Throughput Screen.* **9** 203–12

[13] Sun H, Chattopadhaya S, Wang J and Yao S Q 2006 Recent developments in microarray-based enzyme assays: from functional annotation to substrate/inhibitor fingerprinting *Anal. Bioanal. Chem.* **386** 416–26

[14] Kharati M, Foroutanparsa S, Rabiee M, Salarian R, Rabiee N and Rabiee G 2018 Early diagnosis of multiple sclerosis based on optical and electrochemical biosensors: comprehensive perspective *Curr. Anal. Chem.* **14** 1–11

[15] Ahmadi S, Rabiee N and Rabiee M 2019 Application of aptamer-based hybrid molecules in early diagnosis and treatment of diabetes mellitus: from the concepts towards the future *Curr. Diabetes Rev.* **15** 309–13

[16] Noorirad S N, Pourghasem M, Feizi F, Abedian Z, Ghasemi M, Babazadeh Z and Rabiee N 2018 Time dependent of epigenetic effect of disulfiram on tumor suppressor gene of RASSF1A in Hela cancer cell line *Basic Res. Med. Sci.* **5** 8–13

[17] Reineke U, Volkmer-Engert R and Schneider-Mergener J 2001 Applications of peptide arrays prepared by the SPOT-technology *Curr. Opin. Biotechnol.* **12** 59–64

[18] Li M 2000 Applications of display technology in protein analysis *Nat. Biotechnol.* **18** 1251–6

[19] Rabiee N, Bagherzadeh M and Rabiee M 2019 A perspective to the correlation between brain insulin resistance and Alzheimer: medicinal chemistry approach *Curr. Diabetes Rev.* **15** 255–8

[20] Hajebi S, Rabiee N, Bagherzadeh M, Ahmadi S, Rabiee M, Roghani-Mamaqani H, Tahiri M, Tayebi L and Hamblin M R 2019 Stimulus-responsive polymeric nanogels as smart drug delivery systems *Acta Biomater.* **92** 1–18

[21] Karimi M, Mansouri M R, Rabiee N and Hamblin M 2018 Drug delivery approaches *Advances in Nanomaterials for Drug Delivery: Polymeric, Nanocarbon, and Bio-inspired* vol 1 (San Rafael, CA: Morgan & Claypool)

[22] Lam K S, Salmon S E, Hersh E M, Hruby V J, Kazmierski W M and Knapp R J 1991 A new type of synthetic peptide library for identifying ligand-binding activity *Nature* **354** 82–4

[23] Lam K S, Lebl M and Krchňák V 1997 The 'One-Bead-One-Compound' combinatorial library method *Chem. Rev.* **97** 411–48

[24] Houghten R A 1985 General method for the rapid solid-phase synthesis of large numbers of peptides: specificity of antigen-antibody interaction at the level of individual amino acids *Proc. Natl Acad. Sci. USA* **82** 5131–5

[25] MacBeath G and Schreiber S L 2000 Printing proteins as microarrays for high-throughput function determination *Science* **289** 1760–3

[26] MacBeath G, Koehler A N and Schreiber S L 1999 Printing small molecules as microarrays and detecting protein–ligand interactions en masse *J. Am. Chem. Soc.* **121** 7967–8

[27] Falsey J R, Renil M, Park S, Li S and Lam K S 2001 Peptide and small molecule microarray for high throughput cell adhesion and functional assays *Bioconjug. Chem.* **12** 346–53

[28] Houseman B T, Huh J H, Kron S J and Mrksich M 2002 Peptide chips for the quantitative evaluation of protein kinase activity *Nat. Biotechnol.* **20** 270–4

[29] Houseman B T, Gawalt E S and Mrksich M 2003 Maleimide-functionalized self-assembled monolayers for the preparation of peptide and carbohydrate biochips *Langmuir* **19** 1522–31

[30] Karimi M, Mansouri M R, Rabiee N and Hamblin M 2018 Carbon-based nanomaterials *Advances in Nanomaterials for Drug Delivery* (San Rafael, CA: Morgan & Claypool)

[31] Rabiee N, Kiani M, Bagherzadeh M, Rabiee M and Ahmadi S 2019 *Nanoparticle (NP)-Based Delivery Vehicles* (San Rafael, CA: Morgan & Claypool)

[32] Rabiee N, Rabiee M, Bagherzadeh M and Hamblin M 2019 *Stimuli-Responsive Polymers: Nano-Dimension* (San Rafael, CA: Morgan & Claypool)

[33] Rabiee M, Rabiee N, Salarian R and Rabiee G 2019 *Introduction to Nanomaterials in Medicine* (San Rafael, CA: Morgan & Claypool)

[34] Rabiee M, Rabiee N, Salarian R and Rabiee G 2019 Nanomaterials: concepts *Introduction to Nanomaterials in Medicine* (San Rafael, CA: Morgan & Claypool)

[35] Songyang Z *et al* 1995 Catalytic specificity of protein-tyrosine kinases is critical for selective signalling *Nature* **373** 536–9
[36] Pawson T and Scott J D 1997 Signaling through scaffold, anchoring, and adaptor proteins *Science* **278** 2075–80
[37] Cesareni G 1992 Peptide display on filamentous phage capsids. A new powerful tool to study protein-ligand interaction *FEBS Lett.* **307** 66–70
[38] Tegge W, Frank R, Hofmann F and Dostmann W R G 1995 Determination of cyclic nucleotide-dependent protein kinase substrate specificity by the use of peptide libraries on cellulose paper *Biochemistry* **34** 10569–77
[39] Sills M A, Weiss D, Pham Q, Schweitzer R, Wu X and Wu J J 2002 Comparison of assay technologies for a tyrosine kinase assay generates different results in high throughput screening *J. Biomol. Screen.* **7** 191–214
[40] Rabiee N, Kiani M, Bagherzadeh M, Rabiee M and Ahmadi S *Smart Nanostructures* (San Rafael, CA: Morgan & Claypool)
[41] Vafajoo A *et al* 2018 Multiplexed microarrays based on optically encoded microbeads *Biomed. Microdevices* **20** 66
[42] Rabiee N, Kiani M, Bagherzadeh M, Rabiee M and Ahmadi S 2019 Aptamer-based nano-structures *Nanoparticle (NP)-Based Delivery Vehicles* (San Rafael, CA: Morgan & Claypool)
[43] Rabiee N, Kiani M, Bagherzadeh M, Rabiee M and Ahmadi S 2019 Smart drug/gene delivery systems: toward the future *Nanoparticle (NP)-Based Delivery Vehicles* (San Rafael, CA: Morgan & Claypool)
[44] Deng S J, Bickett D M, Mitchell J L, Lambert M H, Blackburn R K, Carter H L III, Neugebauer J, Pahel G, Weiner M P and Moss M L 2000 Substrate specificity of human collagenase 3 assessed using a phage-displayed peptide library *J. Biol. Chem.* **275** 31422–7
[45] Nazif T and Bogyo M 2001 Global analysis of proteasomal substrate specificity using positional-scanning libraries of covalent inhibitors *Proc. Natl. Acad. Sci.* 98 2967
[46] Turk B E, Huang L L, Piro E T and Cantley L C 2001 Determination of protease cleavage site motifs using mixture-based oriented peptide libraries *Nat. Biotechnol.* **19** 661–7
[47] Turk B E and Cantley L C 2004 Using peptide libraries to identify optimal cleavage motifs for proteolytic enzymes *Methods* **32** 398–405
[48] Meldal M, Svendsen I, Breddam K and Auzanneau F I 1994 Portion-mixing peptide libraries of quenched fluorogenic substrates for complete subsite mapping of endoprotease specificity *Proc. Natl Acad. Sci.* **91** 3314–8
[49] Leon S, Quarrell R and Lowe G 1998 Evaluation of resins for on-bead screening: A study of papain and chymotrypsin specificity using pega-bound combinatorial peptide libraries *Bioorg. Med. Chem. Lett.* **8** 2997–3002
[50] Chen G Y, Uttamchandani M, Zhu Q, Wang G and Yao S Q 2003 Developing a strategy for activity-based detection of enzymes in a protein microarray *Chem. Bio. Chem.* **4** 336–9
[51] Zang X, Yu Z and Chu Y H 1998 Tight-binding streptavidin ligands from a cyclic peptide library *Bioorg. Med. Chem. Lett.* **8** 2327–32
[52] Yu Z, Tu J and Chu Y H 1997 Confirmation of cross-reactivity between Lyme antibody H9724 and human heat shock protein 60 by a combinatorial approach *Anal. Chem.* **69** 4515–8
[53] Chen B, Bestetti G, Day R M and Turner A P 1998 The synthesis and screening of a combinatorial peptide library for affinity ligands for glycosylated haemoglobin *Biosens. Bioelectron.* **13** 779–85

[54] Srebalus C A and Clemmer D E 2001 Assessment of purity and screening of peptide libraries by nested ion mobility-TOFMS: identification of RNase S-protein binders *Anal. Chem.* **73** 424–33

[55] Verdoliva A, Marasco D, De Capua A, Saporito A, Bellofiore P and Manfredi V *et al* 2005 A new ligand for immunoglobulin g subdomains by screening of a synthetic peptide library *Chem. Bio. Chem.* **6** 1242–53

[56] Powell K D and Fitzgerald M C 2004 High-throughput screening assay for the tunable selection of protein ligands *J. Comb. Chem.* **6** 262–9

[57] Rabiee M, Rabiee N, Salarian R and Rabiee G Nanomaterials and biomedical applications *Introduction to Nanomaterials in Medicine* (San Rafael, CA: Morgan & Claypool)

[58] Rabiee M, Rabiee N, Salarian R and Rabiee G 2019 Bioactive nanomaterials *Introduction to Nanomaterials in Medicine* (San Rafael, CA: Morgan & Claypool)

[59] Maghsoudi S *et al* 2019 Recent advancements in aptamer-bioconjugates: sharpening stones for breast and prostate cancers targeting *J. Drug Deliv. Sci. Technol.* **53** 101146

[60] Rabiee N, Kiani M, Bagherzadeh M, Rabiee M and Ahmadi S 2019 Protein-based nanostructures II *Nanoparticle (NP)-Based Delivery Vehicles* (San Rafael, CA: Morgan & Claypool)

[61] Rabiee M, Rabiee N, Salarian R and Rabiee G 2019 Bio-inspired approaches: carbon-based nanomaterials *Introduction to Nanomaterials in Medicine* (San Rafael, CA: Morgan & Claypool) p 11

[62] Nour S, Baheiraei N, Imani R, Rabiee N, Khodaei M and Alizadeh A *et al* 2019 Bioactive materials: a comprehensive review on interactions with biological microenvironment based on the immune response *J. Bionic Eng.* **16** 563–81

[63] Rabiee M, Rabiee N, Salarian R and Rabiee G Porphyrin-based nanomaterials *Introduction to Nanomaterials in Medicine* (San Rafael, CA: Morgan & Claypool)

[64] Karimi M, Mansouri M R, Rabiee N and Hamblin M 2018 Advances in nature-inspired nanomaterials *Advances in Nanomaterials for Drug Delivery: Polymeric, Nanocarbon, and Bio-inspired* (San Rafael, CA: Morgan & Claypool)

[65] Karimi M, Mansouri M R, Rabiee N and Hamblin M 2018 Aptamers and pathogen-based carriers *Advances in Nanomaterials for Drug Delivery: Polymeric, Nanocarbon, and Bio-inspired* (San Rafael, CA: Morgan & Claypool)

[66] Brown P O and Botstein D 1999 Exploring the new world of the genome with DNA microarrays *Nat. Genet.* **21** 33–7

[67] Grant D M 2001 *MSP Technologies for the Analysis of Single Nucleotide Polymorphisms: An Overview* (New York: Marcel Dekker)

[68] Luong D, Kesharwani P, Deshmukh R, Amin M C I M, Gupta U, Greish K and Iyer A K 2016 PEGylated PAMAM dendrimers: enhancing efficacy and mitigating toxicity for effective anticancer drug and gene delivery *Acta Biomater.* **43** 14–29

[69] Lee S, Son S, Song S, Ha T and Choi J 2017 Polyamidoamine (PAMAM) dendrimers modified with cathepsin-B cleavable oligopeptides for enhanced gene delivery *Polymers* **9** 224

[70] Mirminachi B, Mohammadi Z, Merat S, Neishabouri A, Sharifi A H, Alavian S H, Poustchi H and Malekzadeh R 2017 Update on the prevalence of Hepatitis C virus infection among Iranian general population: a systematic review and meta-analysis *Hepat. Mon.* **17** e42291

[71] Delehanty J B and Ligler F S 2002 A microarray immunoassay for simultaneous detection of proteins and bacteria *Anal. Chem.* **74** 5681–7

[72] Council N R 2003 *Materials Research to Meet 21st-Century Defense Needs* (Washington, DC: The National Academies Press) 330 pp
[73] Hanash S 2003 Disease proteomics *Nature* **422** 226–32
[74] Schaeferling M, Schiller S, Paul H, Kruschina M, Pavlickova P, Meerkamp M, Giammasi C and Kambhampati D 2002 Application of self-assembly techniques in the design of biocompatible protein microarray surfaces *Electrophoresis* **23** 3097–105
[75] Pavlickova P *et al* 2002 Antibody detection in human serum using a versatile protein chip platform constructed by applying nanoscale self-assembled architectures on gold *J. Proteome Res.* **1** 227–31
[76] Uchida T, Fukawa A, Uchida M, Fujita K and Saito K 2002 Application of a novel protein biochip technology for detection and identification of rheumatoid arthritis biomarkers in synovial fluid *J. Proteome Res.* **1** 495–9
[77] Flad T, Bogumil R, Tolson J, Schittek B, Garbe C, Deeg M, Mueller C A and Kalbacher H 2002 Detection of dermcidin-derived peptides in sweat by ProteinChip technology *J. Immunol. Methods* **270** 53–62
[78] Batorfi J, Ye B, Mok S C, Cseh I, Berkowitz R S and Fulop V 2003 Protein profiling of complete mole and normal placenta using ProteinChip analysis on laser capture micro-dissected cells *Gynecol. Oncol.* **88** 424–8
[79] Reimer U, Reineke U and Schneider-Mergener J 2002 Peptide arrays: from macro to micro *Curr. Opin. Biotechnol.* **13** 315–20
[80] Fukui S, Feizi T, Galustian C, Lawson A M and Chai W 2002 Oligosaccharide microarrays for high-throughput detection and specificity assignments of carbohydrate-protein interactions *Nat. Biotechnol.* **20** 1011–7
[81] Salisbury C M, Maly D J and Ellman J A 2002 Peptide microarrays for the determination of protease substrate specificity *J. Am. Chem. Soc.* **124** 14868–70
[82] Whiteside T L 2002 Cytokine assays *Biotechniques* **33** 12–5
[83] Howell S W, Inerowicz H D, Regnier F E and Reifenberger R 2003 Patterned protein microarrays for bacterial detection *Langmuir* **19** 436–9
[84] Groves J T 2002 Membrane array technology for drug discovery *Curr. Opin. Drug Discov. Dev.* **5** 606–12
[85] Fang Y, Frutos A G and Lahiri J 2002 Membrane protein microarrays *J. Am. Chem. Soc.* **124** 2394–5
[86] Krieg R C, Paweletz C P, Liotta L A and Petricoin E F 2002 Clinical proteomics for cancer biomarker discovery and therapeutic targeting *Technol. Cancer Res. Treat.* **1** 263–72
[87] Martinez M J, Aragon A D, Rodriguez A L, Weber J M, Timlin J A, Sinclair M B, Haaland D M and Werner-Washburne M 2003 Identification and removal of contaminating fluorescence from commercial and in-house printed DNA microarrays *Nucl. Acids Res.* **31** e18
[88] Stenlund P, Babcock G J, Sodroski J and Myszka D G 2003 Capture and reconstitution of G protein-coupled receptors on a biosensor surface *Anal. Biochem.* **316** 243–50
[89] Pawlak M, Schick E, Bopp M A, Schneider M J, Oroszlan P and Ehrat M 2002 Zeptosens' protein microarrays: a novel high performance microarray platform for low abundance protein analysis *Proteomics* **2** 383–93
[90] Wu G, Datar R H, Hansen K M, Thundat T, Cote R J and Majumdar A 2001 Bioassay of prostate-specific antigen (PSA) using microcantilevers *Nat. Biotechnol.* **19** 856

[91] Shao G, Li D, Zhang J, Yang J and Shangguan Y 2019 Automatic microarray image segmentation with clustering-based algorithms *PLoS One* **14** e0210075

[92] Stich M I, Nagl S, Wolfbeis O S, Henne U and Schaeferling M 2008 A dual luminescent sensor material for simultaneous imaging of pressure and temperature on surfaces *Adv. Funct. Mater.* **18** 1399–406

[93] Nagl S, Schaeferling M and Wolfbeis O S 2005 Fluorescence analysis in microarray technology *Microchim. Acta* **151** 1–21

[94] Smith E A and Corn R M 2003 Surface plasmon resonance imaging as a tool to monitor biomolecular interactions in an array based format *Appl. Spectrosc.* **57** 320A–32A

[95] Häussling L, Ringsdorf H, Schmitt F and Knoll W 1991 Biotin-functionalized self-assembled monolayers on gold: surface plasmon optical studies of specific recognition reactions *Langmuir* **7** 1837–40

[96] Smith E A, Erickson M G, Ulijasz A T, Weisblum B and Corn R M 2003 Surface plasmon resonance imaging of transcription factor proteins: interactions of bacterial response regulators with DNA arrays on gold films *Langmuir* **19** 1486–92

[97] Liebermann T and Knoll W 2000 Surface-plasmon field-enhanced fluorescence spectroscopy *Colloids Surf. A* **171** 115–30

[98] Watson A, Wu X and Bruchez M 2003 Lighting up cells with quantum dots *Biotechniques* **34** 296–300

[99] Yguerabide J and Yguerabide E E 2001 Resonance light scattering particles as ultra-sensitive labels for detection of analytes in a wide range of applications *J. Cell Biochem. Suppl.* **37** 71–81

[100] Ulman A 1996 Formation and structure of self-assembled monolayers *Chem. Rev.* **96** 1533–54

[101] Zammatteo N, Jeanmart L, Hamels S, Courtois S, Louette P, Hevesi L and Remacle J 2000 Comparison between different strategies of covalent attachment of DNA to glass surfaces to build DNA microarrays *Anal. Biochem.* **280** 143–50

[102] Pirrung M C 2002 How to make a DNA chip *Angew. Chem. Int. Ed.* **41** 1276–89

[103] Gun J and Sagiv J 1986 On the formation and structure of self-assembling monolayers: III. Time of formation, solvent retention, and release *J. Colloid Interface Sci.* **112** 457–72

[104] Cohen R, Zenou N, Cahen D and Yitzchaik S 1997 Molecular electronic tuning of Si surfaces *Chem. Phys. Lett.* **279** 270–4

[105] Haab B B, Dunham M J and Brown P O 2001 Protein microarrays for highly parallel detection and quantitation of specific proteins and antibodies in complex solutions *Genome Biol.* **2** 2001–2

[106] Joos T O *et al* 2000 A microarray enzyme-linked immunosorbent assay for autoimmune diagnostics *Electrophoresis* **21** 2641–50

[107] Mendoza L G, McQuary P, Mongan A, Gangadharan R, Brignac S and Eggers M 1999 High-throughput microarray-based enzyme-linked immunosorbent assay (ELISA) *Biotechniques* **27** 778–80

[108] Martin B D, Gaber B P, Patterson C H and Turner D C 1998 Direct protein microarray fabrication using a hydrogel 'Stamper' *Langmuir* **14** 3971–5

[109] Bernard A, Delamarche E, Schmid H, Michel B, Bosshard H R and Biebuyck H 1998 Printing patterns of proteins *Langmuir* **14** 2225–9

[110] Bhatia S K, Teixeira J L, Anderson M, Shriverlake L C, Calvert J M, Georger J H, Hickman J H, Dulcey J J, Schoen C S and Ligler P E 1993 Fabrication of surfaces resistant

to protein adsorption and application to two-dimensional protein patterning *Anal. Biochem.* **208** 197–205

[111] Sreekumar A, Nyati M K, Varambally S, Barrette T R, Ghosh D and Lawrence T S *et al* 2001 Profiling of cancer cells using protein microarrays: discovery of novel radiation-regulated proteins *Cancer Res.* **61** 7585–93

[112] Schweitzer B *et al* 2002 Multiplexed protein profiling on microarrays by rolling-circle amplification *Nat. Biotechnol.* **20** 359–65

[113] Zhu H *et al* 2001 Global analysis of protein activities using proteome chips *Science* **293** 2101–5

[114] Bertilsson L and Liedberg B 1993 Infrared study of thiol monolayer assemblies on gold: preparation, characterization, and functionalization of mixed monolayers *Langmuir* **9** 141–9

[115] Nuzzo R G and Allara D L 1983 Adsorption of bifunctional organic disulfides on gold surfaces *J. Am. Chem. Soc.* **105** 4481–3

[116] Leggett G J, Roberts C J, Williams P M, Davies M C, Jackson D E and Tendler S J B 1993 Approaches to the immobilization of proteins at surfaces for analysis by scanning tunneling microscopy *Langmuir* **9** 2356–62

[117] Mrksich M and Whitesides G M 1997 Using self-assembled monolayers that present oligo (ethylene glycol) groups to control the interactions of proteins with surfaces *Poly(ethylene glycol). ACS Symposium Series* vol 680 (Washington, DC: American Chemical Society) pp 361–73

[118] Willner I I and Katz E 2000 Integration of layered redox proteins and conductive supports for bioelectronic applications *Angew. Chem. Int. Ed. Engl.* **39** 1180–218

[119] Ostuni E, Chapman R G, Liang M N, Meluleni G, Pier G, Ingber D E and Whitesides G M 2001 Self-assembled monolayers that resist the adsorption of proteins and the adhesion of bacterial and mammalian cells *Langmuir* **17** 6336–43

[120] Kambhampati D and Knoll W 2002 Biotin-streptavidin sensor surface: a versatile platform for performing DNA hybridization interactions *Stud. Interface Sci.* **16** 279–316

[121] Riklin A and Willner I 1995 Glucose and acetylcholine sensing multilayer enzyme electrodes of controlled enzyme layer thickness *Anal. Chem.* **67** 4118–26

[122] Jung L S, Nelson K E, Stayton P S and Campbell C T 2000 Binding and dissociation kinetics of wild-type and mutant streptavidins on mixed biotin-containing alkylthiolate monolayers *Langmuir* **16** 9421–32

[123] Gershoni J M and Palade G E 1982 Electrophoretic transfer of proteins from sodium dodecyl sulfate-polyacrylamide gels to a positively charged membrane filter *Anal. Biochem.* **124** 396–405

[124] Mecklenburg M, Danielsson B and Winqvist F 1998 Broad specificity affinity arrays: a qualitative approach to complex sample discrimination WO1997049989

[125] Towbin H, Staehelin T and Gordon J 1979 Electrophoretic transfer of proteins from polyacrylamide gels to nitrocellulose sheets: procedure and some applications *Proc. Natl. Acad. Sci.* **76** 4350–4

[126] Veiseh M, Zareie M H and Zhang M 2002 Highly selective protein patterning on gold–silicon substrates for biosensor applications *Langmuir* **18** 6671–8

[127] Garno J C, Amro N A, Wadu-Mesthrige K and Liu G-Y 2002 Production of periodic arrays of protein nanostructures using particle lithography *Langmuir* **18** 8186–92

[128] Delamarche E, Sundarababu G, Biebuyck H, Michel B, Gerber C, Sigrist H, Wolf H, Ringsdorf H, Xanthopoulos N and Mathieu H J 1996 Immobilization of antibodies on a photoactive self-assembled monolayer on gold *Langmuir* **12** 1997–2006

[129] Morhard F, Pipper J, Dahint R and Grunze M 2000 Immobilization of antibodies in micropatterns for cell detection by optical diffraction *Sens. Actuators B* **70** 232–42

[130] Karakişla M and Saçak M 2000 The preparation of polyaniline/polypyrrole conductive polymer films on polycarbonate-coated Pt electrodes *J. Polym. Sci. Part A: Polym. Chem.* **38** 51–9

[131] Tatsuma T and Watanabe T 1991 Peroxidase model electrodes: heme peptide modified electrodes as reagentless sensors for hydrogen peroxide *Anal. Chem.* **63** 1580–5

[132] Andreescu S, Barthelmebs L and Marty J-L 2002 Immobilization of acetylcholinesterase on screen-printed electrodes: comparative study between three immobilization methods and applications to the detection of organophosphorus insecticides *Anal. Chim. Acta* **464** 171–80

[133] Fici D A, McCormick W, Brown D W, Herrmann J E, Kumar V and Awdeh Z L 2010 A protein multiplex microarray substrate with high sensitivity and specificity *J. Immunol. Methods* **363** 60–6

[134] Johnsson B, Lofas S and Lindquist G 1991 Immobilization of proteins to a carboxymethyldextran-modified gold surface for biospecific interaction analysis in surface plasmon resonance sensors *Anal. Biochem.* **198** 268–77

[135] Dotsey E Y, Gorlani A, Ingale S, Achenbach C J, Forthal D N, Felgner P L and Gach J S 2015 A high throughput protein microarray approach to classify HIV monoclonal antibodies and variant antigens *PLoS One* **10** e0125581

[136] Benters R, Niemeyer C M and Wöhrle D 2001 Dendrimer-activated solid supports for nucleic acid and protein microarrays *Chem. Bio. Chem.* **2** 686–94

[137] de Boer B, Simon H K, Werts M P L, van der Vegte E W and Hadziioannou G 2000 'Living' free radical photopolymerization initiated from surface-grafted iniferter monolayers *Macromolecules* **33** 349–56

[138] Mennig M, Oliveira P W and Schmidt H 1999 Interference coatings on glass based on photopolymerizable nanomer material *Thin Solid Films* **351** 99–102

[139] Schäferling M and Nagl S 2006 Optical technologies for the read out and quality control of DNA and protein microarrays *Anal. Bioanal.Chem.* **385** 500

[140] Sobek J, Bartscherer K, Jacob A, Hoheisel J D and Angenendt P 2006 Microarray technology as a universal tool for high-throughput analysis of biological systems *Comb. Chem. High Throughput Screen.* **9** 365–80

[141] Huang R P, Huang R, Fan Y and Lin Y 2001 Simultaneous detection of multiple cytokines from conditioned media and patient's sera by an antibody-based protein array system *Anal. Biochem.* **294** 55–62

[142] Mark D, Haeberle S, Roth G, von Stetten F and Zengerle R 2010 Microfluidic lab-on-a-chip platforms: requirements, characteristics and applications *Chem. Soc. Rev.* **39** 1153–82

[143] Kessler N, Ferraris O, Palmer K, Marsh W and Steel A 2004 Use of the DNA flow-thru chip, a three-dimensional biochip, for typing and subtyping of influenza viruses *J. Clin. Microbiol.* **42** 2173–85

[144] Wojciechowski J, Danley D, Cooper J, Yazvenko N and Taitt C R 2010 Multiplexed electrochemical detection of Yersinia pestis and staphylococcal enterotoxin B using an antibody microarray *Sensors* **10** 3351–62

[145] Indermuhle P F, Lin P, Tan M X, Zaugg F G, Zwahlen P and Wagner P (ed) 2002 A microfluidic platform for contactless, dropletless dispensing of high-density protein micro-arrays *Micro Total Analysis Systems* (Dordrecht: Springer)

[146] Fixe F, Dufva M, Telleman P and Christensen C B V 2004 Functionalization of poly (methyl methacrylate) (PMMA) as a substrate for DNA microarrays *Nucl. Acids Res.* **32** e9

[147] Limbach P A and Meng Z 2002 Integrating micromachined devices with modern mass spectrometry *Analyst* **127** 693–700

[148] Olanrewaju A, Beaugrand M, Yafia M and Juncker D 2018 Capillary microfluidics in microchannels: from microfluidic networks to capillaric circuits *Lab Chip* **18** 2323–47

[149] Swanson P, Gelbart R, Atlas E, Yang L, Grogan T, Butler W F, Ackley D E and Sheldon E 2000 A fully multiplexed CMOS biochip for DNA analysis *Sens. Actuators B* **64** 22–30

[150] Drummond T G, Hill M G and Barton J K 2003 Electrochemical DNA sensors *Nat. Biotechnol.* **21** 1192–9

[151] Boon E M, Ceres D M, Drummond T G, Hill M G and Barton J K 2000 Mutation detection by electrocatalysis at DNA-modified electrodes *Nat. Biotechnol.* **18** 1096–100

[152] Dijksma M, Kamp B, Hoogvliet J C and van Bennekom W P 2001 Development of an electrochemical immunosensor for direct detection of interferon-γ at the attomolar level *Anal. Chem.* **73** 901–7

[153] Nsabimana A, Lan Y, Du F, Wang C, Zhang W and Xu G 2019 Alkaline phosphatase-based electrochemical sensors for health applications *Anal. Methods* **11** 1996–2006

[154] Ho W O, Athey D and McNeil C J 1995 Amperometric detection of alkaline phosphatase activity at a horseradish peroxidase enzyme electrode based on activated carbon: potential application to electrochemical immunoassay *Biosens. Bioelectron.* **10** 683–91

[155] Zhang L, Zhu C, Geng J, Shi X, Gao Y, Chang Z and Qian H 2015 Silicon-based integrated microarray biochips for biosensing and biodetection applications *Biosensors – Micro and Nanoscale Applications* (Rijeka: InTech)

[156] MacBeath G 2002 Protein microarrays and proteomics *Nat. Genet.* **32** 526–32

Chapter 3

Protein microarrays: materials and fabrication methods

Navid Rabiee, Sakineh Hajebi, Sepideh Ahmadi, Mohammad Rabiee, Mojtaba Bagherzadeh, Saeid Maghsoudi and Michael R Hamblin

3.1 Materials

3.1.1 Microarrays on polymer substrates

Most of the previous discussion in this book has concentrated predominantly on two-dimensional surface architectures. Since the functional properties of a protein are extremely dependent on its structure and conformation, it is possible that sensitive proteins under the harsh conditions of a 2D surface may not be able to survive intact. Other arrangements based on 3D surfaces have been proposed which can supply a fluidic environment that closely mimics the biological environment of proteins. Some prominent examples of 3D-based systems are polymer membranes and hydrogels. The basic fundamental principle of a 3D system is that the fluidic environment maintains the protein conformation and consequently the activity of the protein. Other benefits of 3D architecture are that it allows a higher probe density and a larger surface area. The first commonly used substrates for the preparation of microarrays were polymer membranes such as nitrocellulose or aminated nylon. More than two decades later, these platforms are extensively used in microbiology laboratories as blotting supports. Identification of proteins on these membrane supports is normally by radioimaging techniques, although other identification technologies such as fluorescence and chemiluminescence are becoming common [1–4].

Most proteomic investigations using gel-based studies are combined with downstream operations such as mass spectrometry. Hence it is possible to transfer antibodies and proteins after separation by electrophoresis directly onto the blotting membranes. For the past decade, the field of real-time protein interaction studies has been dominated by gold substrates in hydrogel-based arrays. Due to the

doi:10.1088/978-0-7503-3667-3ch3

extensive application of SPR technology by Biacore, this has become increasingly possible [5–8].

3.1.2 Optical resins

Optical resins are used in biomedical applications according to the selection guide for plastics used in healthcare applications. Optical resins are plastic substrates which are appropriate for protein and other microarrays because of their advantages including: biocompatibility, optical properties, surface properties, mechanical and chemical stability, and good rheological properties [9].

3.1.3 Silicon

Silicon is a common material used as a base for microarrays. Figure 3.1 lists the more important companies which have developed silicon chips on a micro to nano scale. The process of preparing silicon chips is relatively simple using printing and lithography. The lithography includes three main steps: (1) deposition, (2) lithography and (3) etching.

3.1.4 Microarrays based on glass substrates

The most popular chip surface for protein microarrays analysis of protein interactions is still glass slides that have long been used for DNA arrays, based on different surfaces and formats. Glass slides coated by polylysine or epoxy-alkylsilanes or by aldehyde-terminated alkylsilanes can also be used to immobilize proteins, antigens, and antibodies. These activated glass slides (slideshow microscope) in each density and format can be viewed using the fluorescence microscope by the end-user. The identification technique in such systems is essentially based on optical techniques such as fluorescence [10, 11].

The reactive coatings on the glass surface are prepared according to self-assembly methods. Self-assembled monolayers (SAMs) of alkylalkoxysilanes or alkylchlorosilanes

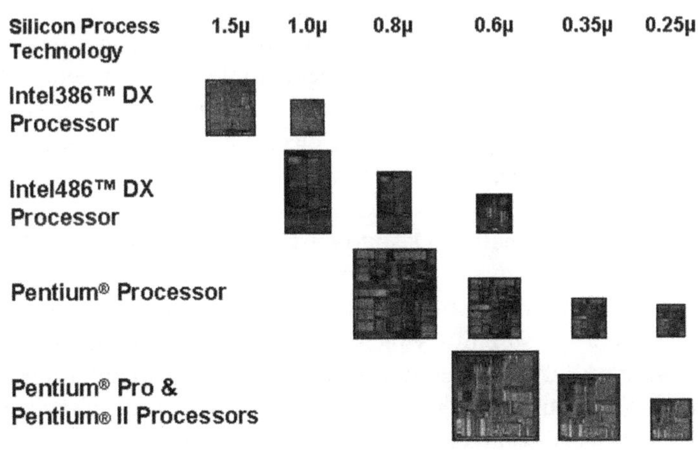

Figure 3.1. Some companies with the technology to produce microscale silicon chips.

bind to hydroxylated surfaces as substrates for their formation. The driving force for this self-assembly is the reaction to form polysiloxane, which is coupled to surface silanol groups (SiOH) via Si–O–Si bonds. The SAMs protect the proteins from structural change and denaturation, and the biocompatibility of the surface increases by the addition of thin organic films [12–14].

3.1.5 Microarrays based on gold substrates

The detection of protein interactions by using gold-based substrates has a rather long history. Thin gold films (5–200 nm) can easily be deposited onto solid supports analogous to silicon or glass wafers, instead of using polycrystalline gold surfaces by sputtering or evaporation in an ultrahigh vacuum (UHV). Before the gold can be deposited, a pre-coating with a thin layer of Ti A-10 nm or Cr is required due to the poor adhesion of gold on native silicon or glass surfaces [15–17].

In the mid-1980s, the spontaneous adsorption of organosulfur compounds onto gold became well known due to publications of Bain and Whitesides [18] and Nuzzo and Allara [19]. This technique has been extensively used for the preparation of self-assembled monolayers (SAMs). If the carbon chain of the alkane thiol has the appropriate length, the resulting molecular construction is extremely stable and orients itself closely vertical to the surface. The nearby neighborhood of the absorbed molecules allows van der Waals forces to play an important role in the inter-molecular stabilization. The thickness of these monolayers is typically in the nanometer range in the perpendicular orientation. It should be mentioned that these monolayers can extend to the whole macroscopic length of the horizontal plane, i.e., more than the surface dimension of the substrate.

The monolayers prepared from alkanethiols on gold have been explored widely (despite SAMs having been fabricated on a wide range of surfaces) for the following reasons: (A) stability of the chemicals under laboratory conditions; (B) availability of the materials; (C) ease of fabrication; (D) flexibility in introducing diverse chemical functional groups to tailor the surface properties; and (E) ability to be constructed without any flaws. They enable exceptional control of the surface properties by engineering and tailoring the structural blocks. Modified surfaces with good biocompatibility which can meet the complex requirements of proteomic studies can be produced. SAMs on gold have the added benefit of surface features that can be characterized by mechanical, electrochemical, and optical analytical tools, unlike glass or polymer-based arrays [20–22].

Many commercial suppliers have started to offer gold-coated microscope slides with appropriate surface quality. Despite all the above benefits, gold-based microarrays still are not in extensive use because of their uncertain stability (thermal and chemical) and cost. The weak interactions between sulfur and gold in the SAMs, lead to poor-temperature stability, which is one of the main constraints on their use. Microarrays were found not to be appropriate for DNA array applications where high-temperature conditions and rigid washing are employed. In the case of protein arrays (where assays are conducted at low temperatures) this is not such a concern.

These surfaces can be used in more advanced mass spectrometry and label-free detection methods such as SPR [23–25].

3.2 Methods

The process of preparation of a silicon chip using lithography, as a substrate for a protein or DNA microarray or for a microfluidic chip is briefly explained in this section. Photolithography (or 'optical lithography') is a process used in microfabrication to selectively remove parts of a thin film or the bulk of a substrate. It uses light to transfer a geometric pattern from a photo-mask to a light-sensitive chemical 'photoresist', or simply 'resist,' onto the surface of the substrate. A series of chemical treatments then either engrave the exposure pattern into the substrate, or enable deposition of a new material in the desired pattern upon the substrate underneath the photoresist. In complex integrated circuits, for example a modern CMOS chip, a wafer will go through the photolithographic cycle up to 50 times [26–28]. Photolithography shares some fundamental principles with photography in that the pattern in the etching resist is created by exposing it to light, either directly (without using a mask) or with a projected image using an optical mask. In etching, a liquid ('wet') or plasma ('dry') chemical agent removes the uppermost layer of the substrate in the areas that are not protected by the photoresist. This procedure is comparable to a high precision version of the method used to make printed circuit boards. Subsequent stages in the process have more in common with etching than with lithographic printing. Photolithography is used because it can create extremely small patterns (down to a few tens of nanometers in size), and affords exact control over the shape and size of the objects it creates. Moreover, it can create patterns over an entire surface in a cost-effective manner. The main disadvantages are that it requires a flat substrate to start with, it is not very effective at creating shapes that are not flat, and it can require extremely clean operating conditions [29–32].

3.2.1 Design

High-throughput ability is related to a sufficient spot density, i.e., the capability to immobilize a definite number of spots with diverse biomolecules (haptens, proteins, or peptides) which function as capture probes that attach to target molecules in the fluidic sample and localize them onto a finite area. Technological solutions exist to produce high spot densities such as pin, piezo-electric, or pin and ring arrayers. The use of proteomic analysis means that very complex biological solutions must interact efficiently with smart interface designs. To counter this, the array surface design must meet some requirements similar to those of different biosensors:

- The ability to retain immobilized biomolecules in an active state.
- Low capacity for nonspecific absorption.
- High immobilized density of the receptors.
- Ability to differentiate between biologically relevant binding and unwanted physical adsorption of sample material.

Various immobilization strategies and a diversity of functionalized surfaces have been used to date in microarray formats to attach proteins to surfaces. Recent instances can be found in the work of MacBeath *et al*, Brown *et al*, and Snyder *et al.* MacBeath *et al* investigated an approach based on the classical physical absorption of proteins onto a surface, and afterwards blocked with milk or albumin powder as is done in ELISA [33–36].

However, in order to prevent the problem of steric shielding of small proteins and peptides during the BSA blocking, attempts have been made to improve the method by covalently immobilizing BSA on an activated surface and subsequently covalently attaching proteins and peptides onto the upper side of the BSA layer. A microarray with 10 800 spots was prepared using this immobilization strategy. One spot was the FKBP12-rapamycin binding domain (FRB) of FKBP-rapamycin-associated protein, which interacts specifically with human immunophilin FKBP12 in the presence of rapamycin. The remaining 10 799 spots were prepared with protein G which interacts specifically with immunoglobulin G (IgG). Afterwards, this microarray was exposed to a mixture of Cy5-FKBP12 and BODIPY-FL-IgG, with 100 nM rapamycin contained in the buffer. The single FRB spot with Cy5-FKBP12 bound could be clearly recognized among the whole 10 800 spots, 10 799 of them had specifically bound BODIPY-FL-IgG. The test highlighted the potential of protein microarrays [37–40].

3.2.2 Chemical immobilization techniques for proteins

The functional coatings can be constructed via the chemical linkage of proteins by covalent binding to a biocompatible interface, or a microarray substrate. Attachment can also done be via supramolecular interactions or complex coordination chemistry. Moreover electrostatic interactions can attach proteins to the surface. Membranes made from nylon, nitrocellulose, poly-L-lysine can be used, or aminopropylsilane-coated surfaces that allow simple electrostatic binding of proteins. SAMs based on polyethylene glycols [41] or poly(L-lysine)-*g*-poly(ethylene glycol) [42] are known to interact with proteins and are therefore extensively used in microarrays and biosensors. Approaches for coupling proteins using photochemical and covalent approaches are addressed below.

3.2.3 Covalent chemical coupling

In order for proteins to undergo covalent chemical coupling to microarray substrates, they should have amino acids with reactive groups: $-NH_2$ (lysine), –SH (cysteine), –COOH (glutamic acid, aspartic acid), Ph–OH (tyrosine), imidazole (histidine), –OH (serine). Stable bonds are formed between free peptide amino groups and aldehydes, epoxy compounds or NHS (*N*-hydroxy-succinimide) esters. Carboxyl groups can be activated with EDC (*N*-3-dimethylaminopropyl-*N*-ethyl-carbodiimide) and NHS. The substrate surface can also be equipped with carboxyl groups that can be activated as carbodiimides or NHS esters to allow it to react with free amino groups in proteins [43–45]. Other reagents exist that can react with thiol groups such as vinyl sulfone, pyridyl disulfide, and maleimides. The surface can be modified with epoxy groups to react with hydroxyl groups and with isothiocyanates to react with amino groups [46].

3.2.4 Photochemical cross-linking

Many approaches are available for photochemical cross-linking of biomolecules to functionalized surfaces. For example, biomolecules can be coupled to epoxy- or aldehyde-coated slides using irradiation with UV light to trigger the reaction. Other photoreactive groups are available, of which the azidophenyl group is an outstanding example. Substrates can be coated with azidophenyl-functionalized layers, and then proteins are attached by irradiation with visible or UV light. By changing the substituents on the azidophenyl ring, the excitation wavelength for this reaction can be adjusted [47].

Researchers at the commercial firm Exiqon (Hilden, Germany) have developed another strategy for the photochemical bonding of biomolecules. They used an anthraquinone linker, which was first attached to a variety of solid polymeric supports using irradiation (e.g. polypropylene, polyethylene, PMMA, polystyrene, nylon). The linker can then be conjugated to reactive amino acid groups (carboxylic acids or amines) present on biomolecules or ligands [48–50].

The photoaptamer method is another multipurpose approach for photochemical cross-linking of proteins to surfaces, introduced by SomaLogic (Boulder, CO). Photoaptamers were derived from the PhotoSELEX™ procedure, and are highly specific short RNA sequences that recognize proteins and non-covalently bind to them. In this method (as opposed to other approaches in which the probes are immobilized by photochemical cross-linking), the molecular binding event to the probe is stabilized by a sequential photo-crosslinking reaction. Single-stranded RNA molecules are chemically synthesized which can be tagged at their 5-end with thiol, amine, acrydite or biotin groups, and are called photoaptamers. These groups help the covalent coupling of the photoaptamers to the substrate surface, which can be multi-well plates, activated slides, or beads. Brominated deoxyuridine (BrdU) is substituted for the thymidine base (T) within the RNA sequences. After the proteins are allowed to non-covalently bind to the photoaptamers, they are then covalently linked by irradiation, thus producing a covalent bond to an electron-rich amino acid on the protein. This attachment allows washing, labeling, and reading steps to be performed under the harshest and most stringent conditions necessary to reduce background and improve signal [51].

Another technology called Versalinx Protein Microarray was introduced by Prolinx (Bothall, WA) using a synthetic chemical, which is based on the salicylhydroxamic acid (SHA) with reversible complexation of phenyldiboronic acid (PDBA). In this regard, 3D polymer brushes with SHA as their terminal unit are layered on the microarray surface, which allows the reversible coupling of PDBA-conjugated proteins to be used [52, 53].

3.3 Preparation of protein arrays

The peptide microarray synthesis methods can be divided into two groups: (1) spotting array methods; and (2) photolithographic or SPOT methods using parallel on-chip synthesis. The spotting array includes pre-synthesis of an array of probes and their direct transfer and subsequent immobilization on a substrate composed of a membrane or glass. This approach has been used to produce an array of almost any type of molecule so it is very versatile. Parallel on-chip synthesis produces peptides on the chip surface using *in situ* peptide synthesis [54].

3.3.1 Light-directed parallel chemical synthesis

The first study on peptide arrays was carried out by the Fodor group [55]. They designed a new technology that concurrently produced thousands of peptides on a glass slide that were spatially addressable by light using a combination of photolithography and solid-phase chemistry.

Another example produced a varied collection of peptides by chemical synthesis. The N-terminal amino acids were protected by NVOC (6-nitroveratryloxycarbonyl), which is a photolabile group (removable by light) that allowed the sequential building of peptides on a glass slide using solid peptide synthesis [56]. A set of photomasks carefully controlled each cycle of the peptide synthesis permitting selective deprotection of the N-terminal amino groups for the growing peptide chain, preventing any unwanted errors. Selective coupling is another technique which is used to both specificity and high affinity using a device on a chip, which enables antibodies attaching for specific peptide sequences. For the first time, it was demonstrated that thousands of μm-size spots of molecules within a small measurement can be produced, and this allows high-throughput screenings of biological experiments and potential miniaturization. The combination of semiconductor and chemistry has made a basic science in biotechnology for the revolution of DNA microarrays. About two years later, some evidence regarding the manufacture of other types of arrays compound of peptidomimetic unnatural biopolymers and oligonucleotides was reported [57].

Unfortunately, these techniques were not as successful as had been expected, due to technical difficulties in providing sufficient masks. If 20 specific photolabile protected amino acids could be used in each elongation cycle to build decamer peptides, it would require $20 \times 10 = 200$ masks On the other hand, DNA microarrays are much easier because there are only four nucleotide bases so for each coupling cycle, only four masks are needed. That is why it is more difficult to produce high-quality peptides on a glass slide, compared to oligonucleotide synthesis. There are also a number of modified on-chip synthetic strategies. Gao *et al* also demonstrated a maskless microarray similar to oligonucleotide arrays, and produced peptides by using photolabile amino acids and digital photolithography [58]. This method was based on a custom-made peptide synthesizer, which uses a computer-controllable optical module, providing a powerful photolithographic technique more affordable in terms of price for most research laboratories.

Another recent photolithographic synthesis technique was based on the production of an array of peptoids (peptide analogs) using a single photolabile synthon, which is an oligomer of N-substituted glycine residues [59]. Kodadek and colleagues used a light-activated procedure that leads to the production of peptides by the 'submonomer' pathway. This process included a four-step cycle for each stage that could be used to synthesize peptoids. First, glycolic acid was protected by a light-sensitive MeNPOC group and attached to an amine-modified surface. Then the MeNPOC group was removed using UV irradiation and the resulting hydroxyl group was activated with tosyl chloride. Eventually, the tosylate was replaced with an amine for the formation of the monomer step. This technique permits the spatially addressable synthesis of peptides onto an array by photolithography, since the hydroxyl group activation, unmasking, and amine replacement will only occur at addresses that have been exposed to UV light. The downside of these photochemical approaches is that the chemistry is much more confined. Difficulties with lack of purity, instability, etc means that only small peptides can be prepared [60].

3.3.2 SPOT™ synthesis

In 1990, Frank *et al* introduced a method called SPOT synthesis for the concurrent synthesis of parallel peptides onto a membrane support. That instrument has become very popular to study aspects of molecular diagnostics [61]. In recent years, techniques like synthesis kits and custom-made membrane-bound peptides have become commercially available, and are a good choice due to their reliability. In 1993 another technique called 'semi-automated SPOT synthesizer' was introduced by the company ABIOMED (Aachen, Germany), which enabled rapid synthesis of many peptides on paper or membrane supports. This method could use other coupling reagents and Fmoc-protected amino acids, and subsequent deprotection could create the target peptide using conventional coupling steps. Many new techniques and assays based on similar formats have been developed using peptide arrays for an accurate analysis of proteins and polypeptides. In conclusion, the advantages of using SPOT methods are: (1) good compatibility of this strategy with an extensive range of screening methods; (2) availability of efficient methods for the production of arrays [62, 63]. The SPOT synthesis method has been improved for better analysis of various classes of proteins; the disadvantage here is that the density of the spots created by this technique is rather low (e.g. 25 spots per cm^2) making high throughput applications difficult.

3.3.3 Spotting microarray approach

This is a supplementary approach to the SPOT technique, creating a better platform for performing enzymatic assays and epitope-mapping. Today, the spotting array format with the new advances in surface modification technologies and in micro-fabrication is becoming a popular tool for high-throughput screening. Another advantage is the use of an automatic arrayer for high-throughput spotting of pre-synthesized peptides on a solid surface. Another factor that makes this a much more effective approach, is that the probe compound has to be synthesized only once, and many repeats can be simply generated in additional spots [64].

The development of more efficient solid supports, better linkers, new peptide-binding chemistries, and automatic synthesis systems means that peptide synthesis has become much easier. Peptide libraries can be arrayed onto supports using procedures such as mixed synthesis, split parallel synthesis, reagent mixture synthesis, among others. It is also very important to choose an appropriate immobilization surface and suitable functional groups to allow the attachment of proteins or peptides to be effective. Common immobilization surfaces such as nitrocellulose membranes, polyvinylidene fluoride, polystyrene, thin film agarose etc, would likely not be suitable for a microarray template, due to the fact that these surfaces use non-covalent forces such as hydrophobic interactions for binding, and moreover result in the production of low-density arrays of randomly assorted biomolecules. As a result, these surfaces frequently demonstrate a comparatively low signal-to-noise ratio in peptide/protein screening applications. However, glass slides could be the ideal surface in as much as they are quite cheap, have low inherent fluorescence, and a homogeneous chemical surface that can provide extremely high density arrays of biomolecules. The important features of an immobilization technique include the following: (1) before immobilization, the approach should require only minor post-synthetic modification of ligands; (2) allow lower concentrations of reagents to react rapidly; (3) the immobilization procedure must take advantage of functional groups already existing in peptides, such as carboxylic acids, thiols, amines, alcohols, etc. Furthermore, the process must allow conventional spectroscopic methods to be used for read-out, and have well-behaved kinetics. Immobilization chemistries possessing one or more of the above-mentioned features have been described by several groups [65, 66].

One approach introduced by the Schreiber group, successfully generated protein and peptide high-density microarrays that could be spotted with an automatic robotic arm. That meant that simultaneous screening of tens of thousands of peptides immobilized on a glass surface was now possible. In contrast to the Fodor method and SPOT technology, Schreiber's technique had advantages such as its ease of use and its ability to be used with almost any type of biomolecules. The authors deposited a molecular layer of BSA onto the glass surface, followed by derivatization of the BSA with *N*-hydroxysuccimide (NHS) which was attached to the peptides resulting in only slight nonspecific protein absorption, and the ideal spatial orientation to allow the maximum interaction between ligands and proteins, etc. The spotting of the peptides was carried out by covalent reaction between the nucleophilic groups in the peptides (–SH, –NH2, –OH, etc) and the reactive NHS group on the BSA surface. The limitation of this approach was that non-site-specific immobilization might cause difficulties if simultaneous generation of a high-density array of peptides which retained their biological activity was required [67, 68].

It is imperative for peptides to be immobilized with the correct orientation in order to be able to interact efficiently with their targeting proteins, despite not possessing a well-defined 3D structure. Falsey *et al* [69] described an approach for regiospecific immobilization of peptides on a glass slide using N-terminal cysteine-containing peptides immobilized on glyoxylic acid functionalized slides, by thiazolidine ring formation or oxime bond formation. Commercially available slides were constructed using APETS in order to create amine-functionalized slides which were then converted to glyoxyl derivative glass slides via two steps of: (1) linkage of Fmoc-Ser followed by oxidation and deprotection, and (2) linkage to protected glycolic acid and deprotection. Despite being the first description of chemoselective ligation of peptides on a glass surface, the limitations of this method were the comparative instability of the oxime bond, and difficult thiazolidine ring formation due to the unfavorably restricted orientation, which delayed the reaction of the peptides with their target proteins. These were the main obstacle to using this strategy. Mrksich and coworkers used the Diels–Alder reaction for attaching N-terminal cyclopentadiene-containing peptides onto the SAM surface bearing quinone groups (figure 3.2) [70].

Peptides that had been functionalized with a cyclopentadiene group were used to attach to a self-assembled monolayer of alkanethiols derivatized with benzoquinone groups. The Diels–Alder reaction between the quinone groups and the cyclodiene allowed rapid and optional immobilization of the peptides onto the glass. This technique had many benefits over other alternative approaches. There was no need for blocking, as the self-assembled monolayers were inactive and allowed no non-specific reactions with proteins. Also, the immobilized peptide ligands were in an orderly homogeneous array allowing more quantitative assays. However, the need to conjugate an unusual cyclopentadiene group to the peptides was synthetically challenging [71].

The use of maleimide-terminated self-assembled monolayers for the immobilization of peptides and carbohydrates was later proposed [72]. In this method the reaction between maleimide and thiol groups on a gold substrate was used for immobilization of ligands in self-assembled monolayers of alkanethiolates requiring no post-synthetic modification of ligands. Monitoring of the real-time interaction between proteins or carbohydrates and the immobilized peptides was performed by SPR spectroscopy. Reddy and Kodadek used a similar approach to construct peptoid arrays to provide a 'molecular fingerprint' with binding to three fluorescently tagged proteins, i.e., ubiquitin, GST, and maltose-binding protein [73]. Immobilization of octameric peptoids by their C-terminal cysteines onto maleimide-functionalized glass slides was successfully achieved allowing screening of protein fingerprints. This approach showed that when assayed in the presence of other bacterial proteins, the exclusive fingerprint of GST could be distinguished.

Another method for site-specific immobilization of peptides on a glass slide has been described to overcome limitations like unstable peptide attachment, and the 'orientation and effective interaction' problem in peptide microarrays. This used N-terminal cysteine residues on peptide substrates that were attached to thioester moieties on the array surface. The most notable benefits of this approach were: (1) peptides containing N-terminal cysteine can be easily synthesized using standard peptide synthesis methods with no need for additional modification; (2) creating a thioester-derivatized glass surface is also easy; (3) the chemical reaction thiol

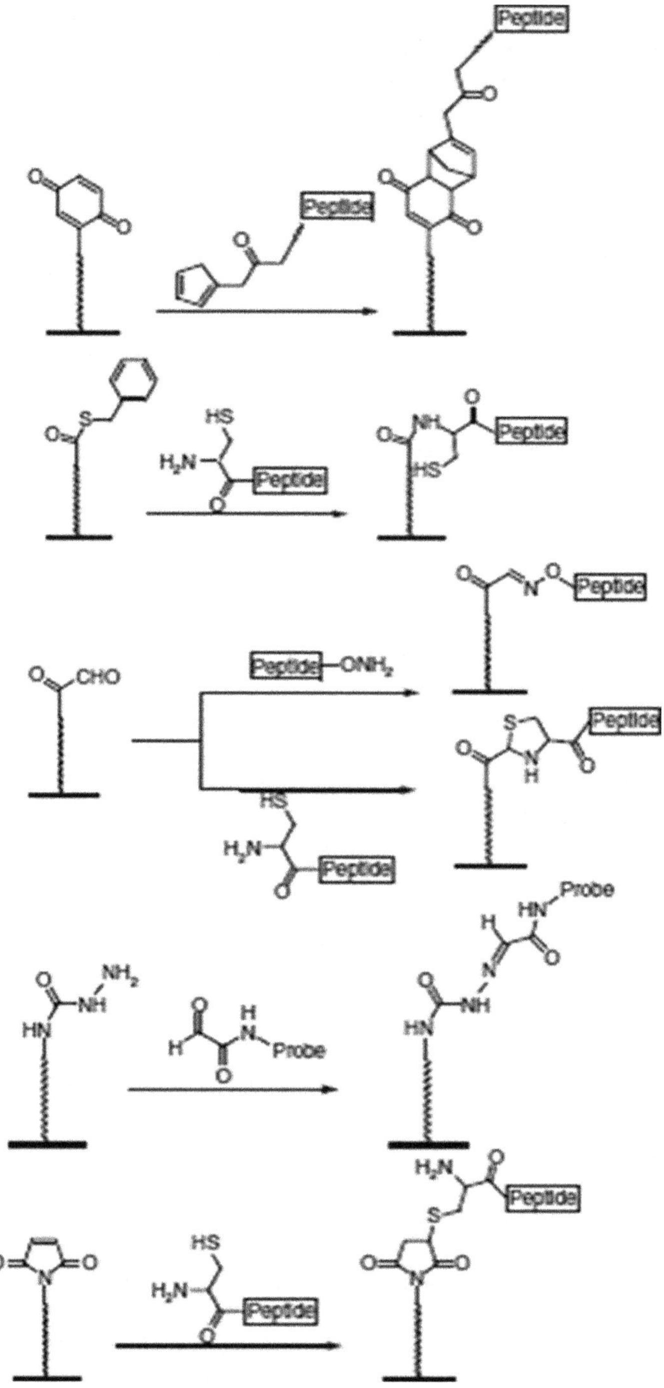

Figure 3.2. Chemoselective immobilization chemistries in peptide microarrays.

peptides and thioesters takes place under physiological conditions and there is no need for additional reagents [74, 75].

Another method used immobilization of a glyoxylyl peptide that reacted with a semicarbazide functionalized glass slide. Using a synthetic glyoxylyl fluorescent probe allowed the optimization of the homogeneity and functional density possible. Probes that were tagged with biotin could be attached to the semicarbazide slides and allowing recognition with an anti-biotin antibody or streptavidin.

Recently, Salisbury and his colleagues reported an effective way to screen protease substrate specificity using peptide microarrays [76]. A library containing 361 different spatially addressable peptides was immobilized on aldehyde-derivatized glass slides. These peptides containing potential protease substrate sequences were conjugated with the fluorescent tag 7-amino-4-carbamoylmethyl coumarin (ACC). Proteases preferentially cleaved specific substrate sequences, and as the peptide was being released thus unmasking the ACC fluorescence that could be measured. Using an analogous method, our group (figure 3.3) used a derivative of fluorogenic coumarin to create a series of substrates that could be susceptible to hydrolysis by different enzymes, and were immobilized on a glass slide to generate a small molecule-based microarray with the ability to sensitively detect various hydrolytic enzymes [77]. Every conjugate included two separate components: an enzyme susceptible grouping and a fluorogenic moiety. The fluorogenic group functioned as a sensitive reporter group for enzymatic activities giving a fluorescence read-out. The bifunctional coumarin derivative possessed a carboxyl group for immobilization on a glass surface, and an electron-donating group (anilide or phenolic group) that could be used as the site for conjugation to a potential enzyme substrate.

The potential enzyme substrate could be tailored in order to detect various enzymes of choice. As enzymatic cleavage occurred, the highly fluorescent coumarin was released from the surface of the glass slide. The inspiration for our approach was originally described by Goddard *et al* who showed that similar methods could be used in a microplate arrangement to produce ‘fingerprints’ of a diversity of enzymes [78].

The Staudinger ligation is an approach for site-specific immobilization of peptides onto a microarray. An azido group is attached to either the main chain or to a side chain in the peptide, which then reacts with a phosphinothioester-derivatized surface to create an amide bond which is a rapid high-yield reaction occurring at room temperature in both aqueous and wet organic solvents. Kiyonaka and his colleagues described supramolecular hydrogels which provided a semi-wet environment that was more compatible with enzyme activity [79]. Aqueous cavities were formed in the gel matrix that provided a semi-wet environment for enzyme reactions, while the hydrophobic regions of the fiber could provide a site for monitoring the reaction by fluorescent dyes.

Cheng and his colleagues [80] described a new method to prepare chips by using a metal-chelating hexapeptide allowing attachment of peptides by addition of a specific metal ion. This method could be used to purify histidine-tagged proteins or other targets that possessed a histidine label. Marcella *et al* used a glass slide

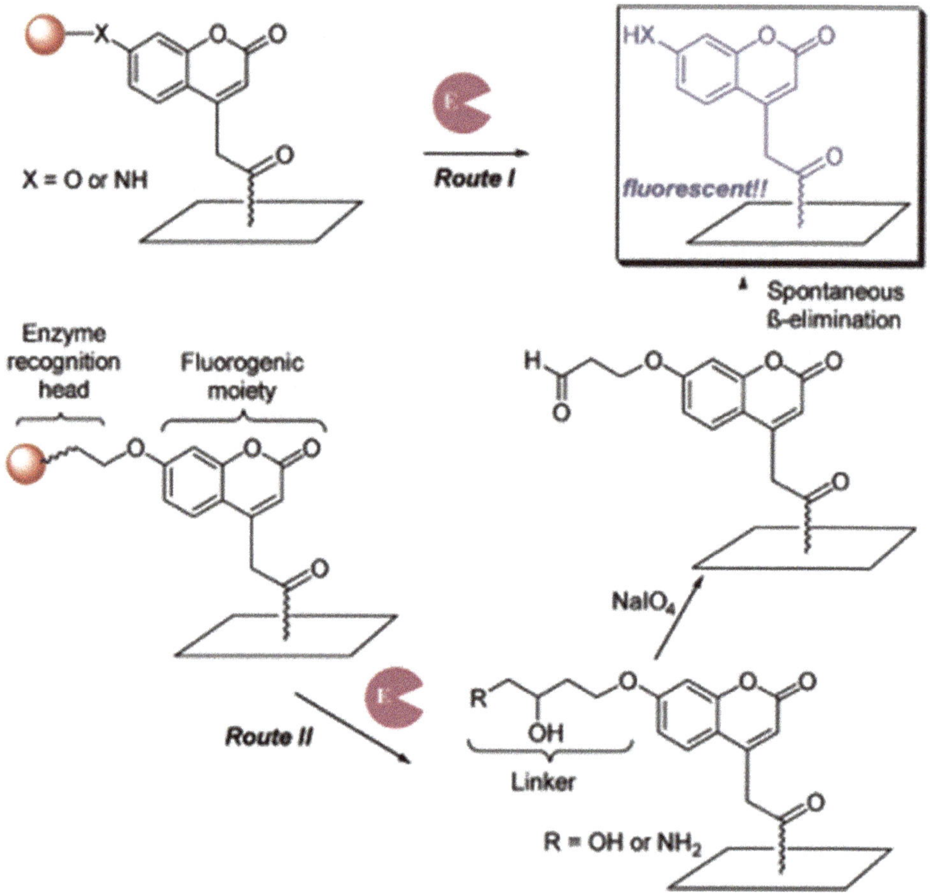

Figure 3.3. Approaches for recognition of hydrolytic enzymes. Reprinted with permission from [77]. Copyright (2020) American Chemical Society.

covered with a copolymer of NAS, DMA, and MAPS that could be used as a peptide array substrate [81]. The peptide immobilization was accomplished onto a polymeric surface. Andresen and coworkers recently described a different method for constructing practical peptide microarrays involving the solution-phase binding of biotinylated synthetic peptides to NeutrAvidin (NA) followed by microdispensing of peptide-NA-complexes onto activated glass surfaces. Antibodies formed links between surface-immobilized peptide probes and fluorescence-labeled secondary antibodies. By selecting spacer molecules of diverse type and length for the NA-modified peptides, the authors found that the spacer length governed the direction of peptide immobilization to provide efficient antibody binding [82]. As opposed to an *in situ* synthesis technique, the spotting microarray method could produce several copies of the identical chip with the help of automatic spotting methods and efficient immobilization methods. In recent years an increasing number of studies have addressed immobilization of peptides on microarrays, but it is important to continue to explore more new and efficient methodologies [83].

References

[1] Nasseri B, Soleimani N, Rabiee N, Kalbasi A, Karimi M and Hamblin M R 2018 Point-of-care microfluidic devices for pathogen detection *Biosens. Bioelectron.* **117** 112–28

[2] Farjadian F, Moghoofei M, Mirkiani S, Ghasemi A, Rabiee N, Hadifar S, Beyzavi A, Karimi M and Hamblin M R 2018 Bacterial components as naturally inspired nano-carriers for drug/gene delivery and immunization: Set the bugs to work? *Biotechnol. Adv.* **36** 968–85

[3] Parsa S F *et al* 2018 Early diagnosis of disease using microbead array technology: a review *Anal. Chim. Acta* **1032** 1–17

[4] Ghasemi A, Rabiee N, Ahmadi S, Hashemzadeh S, Lolasi F and Bozorgomid M *et al* 2018 Optical assays based on colloidal inorganic nanoparticles *Analyst* **143** 3249–83

[5] Vafajoo A *et al* 2018 Multiplexed microarrays based on optically encoded microbeads *Biomed. Microdevices* **20** 66

[6] Rabiee N, Safarkhani M and Rabiee M 2018 Ultra-sensitive electrochemical on-line determination of Clarithromycin based on Poly (L-Aspartic acid)/graphite oxide/pristine graphene/glassy carbon electrode *Asian J. Nanosci. Mater.* **1** 63–73

[7] Deljoo S, Rabiee N and Rabiee M 2019 Curcumin-hybrid nanoparticles in drug delivery system *Asian J. Nanosci. Mater.* **2** 66–91

[8] Karimi M, Mansouri M R, Rabiee N and Hamblin M R 2018 *Advances in Nanomaterials for Drug Delivery: Polymeric, Nanocarbon, and Bio-inspired* (San Rafael, CA: Morgan & Claypool)

[9] Vafajoo A, Salarian R and Rabiee N 2018 Biofunctionalized microbead arrays for early diagnosis of breast cancer *Biomed. Phys. Eng. Express* **4** 065028

[10] Kharati M, Foroutanparsa S, Rabiee M, Salarian R, Rabiee N and Rabiee G 2018 Early diagnosis of multiple sclerosis based on optical and electrochemical biosensors: comprehensive perspective *Curr. Anal. Chem.* **14** 1–11

[11] Ahmadi S, Rabiee N and Rabiee M 2019 Application of aptamer-based hybrid molecules in early diagnosis and treatment of diabetes mellitus: from the concepts towards the future *Curr. Diabetes Rev.* **15** 309–13

[12] Noorirad S N, Pourghasem M, Feizi F, Abedian Z, Ghasemi M, Babazadeh Z and Rabiee N 2018 Time dependent of epigenetic effect of disulfiram on tumor suppressor gene of RASSF1A in Hela cancer cell line *Basic Res. Med. Sci.* **5** 8–13

[13] Rabiee N, Bagherzadeh M and Rabiee M 2019 A perspective to the correlation between Brain insulin resistance and Alzheimer: medicinal chemistry approach *Curr. Diabetes Rev.* **15** 255–8

[14] Hajebi S, Rabiee N, Bagherzadeh M, Ahmadi S, Rabiee M, Roghani-Mamaqani H, Tahriri M, Tayebi L and Hamblin M R 2019 Stimulus-responsive polymeric nanogels as smart drug delivery systems *Acta Biomater.* **92** 1–18

[15] Karimi M, Mansouri M R, Rabiee N and Hamblin M R 2018 Drug delivery approaches *Advances in Nanomaterials for Drug Delivery: Polymeric, Nanocarbon, and Bio-inspired* vol 1 (San Rafael, CA: Morgan & Claypool)

[16] Karimi M, Mansouri M R, Rabiee N and Hamblin M R 2018 Carbon-based nanomaterials *Advances in Nanomaterials for Drug Delivery* (San Rafael, CA: Morgan & Claypool)

[17] Rabiee N, Kiani M, Bagherzadeh M, Rabiee M and Ahmadi S 2019 *Nanoparticle (NP)-Based Delivery Vehicles* (San Rafael, CA: Morgan & Claypool)

[18] Bain C D and Whitesides G M 1988 Molecular-level control over surface order in self-assembled monolayer films of thiols on gold *Science* **240** 62–3

[19] Nuzzo R G and Allara D L 1983 Adsorption of bifunctional organic disulfides on gold surfaces *J. Am. Chem. Soc.* **105** 4481–3

[20] Rabiee N, Rabiee M, Bagherzadeh M and Hamblin M R 2019 *Stimuli-Responsive Polymers: Nano-Dimension* (San Rafael, CA: Morgan & Claypool)

[21] Nasab S G, Yazd M J, Semnani A, Kahkesh H, Rabiee N and Rabiee M 2019 Natural corrosion inhibitors *Synthesis Lectures on Mechanical Engineering* vol 14 (San Rafael, CA: Morgan & Claypool) pp 1–96

[22] Bahrami S, Baheiraei N, Mohseni M, Razavi M, Ghaderi A, Azizi B, Rabiee N and Karimi M 2019 Three-dimensional graphene foam as a conductive scaffold for cardiac tissue engineering *J. Biomater. Appl.* **34** 75–85

[23] Rabiee M, Rabiee N, Salarian R and Rabiee G 2019 *Introduction to Nanomaterials in Medicine* (San Rafael, CA: Morgan & Claypool)

[24] Rabiee N, Safarkhani M and Amini M M 2019 Investigating the structural chemistry of organotin (IV) compounds: recent advances *Rev. Inorg. Chem.* **39** 13–45

[25] Rabiee M, Rabiee N, Salarian R and Rabiee G 2019 Nanomaterials: concepts *Introduction to Nanomaterials in Medicine* (San Rafael, CA: Morgan & Claypool)

[26] Rabiee N, Kiani M, Bagherzadeh M, Rabiee M and Ahmadi S 2019 *Smart Nanostructures* (San Rafael, CA: Morgan & Claypool)

[27] Rabiee N, Kiani M, Bagherzadeh M, Rabiee M and Ahmadi S 2019 *An Introduction to Drug/gene Delivery Systems* (San Rafael, CA: Morgan & Claypool)

[28] Rabiee N, Kiani M, Bagherzadeh M, Rabiee M and Ahmadi S 2019 Aptamer-based nanostructures *Nanoparticle (NP)-Based Delivery Vehicles* (San Rafael, CA: Morgan & Claypool)

[29] Rabiee N and Rabiee M 2019 A promising stimuli-responsive nanocomposite as a theranostic agent for targeted delivery *J. Bioeng. Res.* **1** 27–36

[30] Rabiee N, Kiani M, Bagherzadeh M, Rabiee M and Ahmadi S 2019 Smart drug/gene delivery systems: toward the future *Nanoparticle (NP)-Based Delivery Vehicles* (San Rafael, CA: Morgan & Claypool)

[31] Rabiee, Rabiee N, Salarian R and Rabiee G 2019 Nanomaterials and biomedical applications *Introduction to Nanomaterials in Medicine* (San Rafael, CA: Morgan & Claypool)

[32] Rabiee M, Rabiee N, Salarian R and Rabiee G 2019 Bioactive nanomaterials *Introduction to Nanomaterials in Medicine* (San Rafael, CA: Morgan & Claypool)

[33] Maghsoudi S *et al* 2019 Recent advancements in aptamer-bioconjugates: Sharpening Stones for breast and prostate cancers targeting *J. Drug Deliv. Sci. Technol.* **53** 101146

[34] Rabiee N, Kiani M, Bagherzadeh M, Rabiee M and Ahmadi S 2019 Protein-based nanostructures II *Nanoparticle (NP)-Based Delivery Vehicles* (San Rafael, CA: Morgan & Claypool)

[35] Moshayedi H R, Rabiee M and Rabiee N 2019 A novel graphene-based nanosensor for detection of ethanol gas *Iranian J. Sci. Technol. Trans. A: Sci.* **43** 2227–37

[36] Rabiee M, Rabiee N, Salarian R and Rabiee G 2019 Bio-inspired approaches: carbon-based nanomaterials *Introduction to Nanomaterials in Medicine* vol 8 (San Rafael, CA: Morgan & Claypool) p 11

[37] Rabiee N and Rabiee M 2018 Magnetic stimuli-responsive cobalt ferrite nanoparticle as theranostic agents for targeted delivery *Curr. Nanomater.* **3** 160–7

[38] Nour S, Baheiraei N, Imani R, Rabiee N, Khodaei M, Alizadeh A and Moazzeni S M 2019 Bioactive materials: a comprehensive review on interactions with biological microenvironment based on the immune response *J. Bionic Eng.* **16** 563–81

[39] Rabiee M, Rabiee N, Salarian R and Rabiee G 2019 Enzyme-responsive nanomaterials *Introduction to Nanomaterials in Medicine* (San Rafael, CA: Morgan & Claypool)

[40] Chavoshi S, Rabiee M, Rafizadeh M, Rabiee N, Shamsabadi A S, Bagherzadeh M, Salarian R, Tahiri M and Tayebi L 2019 Mathematical modeling of drug release from biodegradable polymeric microneedles *Bio-Design Manuf.* 2 96–107

[41] Frutos A G, Brockman J M and Corn R M 2000 Reversible protection and reactive patterning of amine- and hydroxyl-terminated self-assembled monolayers on gold surfaces for the fabrication of biopolymer arrays *Langmuir* **16** 2192–7

[42] Ruiz-Taylor L A, Martin T L, Zaugg F G, Witte K, Indermuhle P, Nock S and Wagner P 2001 Monolayers of derivatized poly(L-lysine)-grafted poly(ethylene glycol) on metal oxides as a class of biomolecular interfaces *Proc. Natl. Acad. Sci.* 98 852–7

[43] Rabiee M, Rabiee N, Salarian R and Rabiee G 2019 Porphyrin-based nanomaterials *Introduction to Nanomaterials in Medicine* (San Rafael, CA: Morgan & Claypool)

[44] Nasseri B, Kocum I C, Seymen C M and Rabiee N 2019 Penetration depth in nanoparticles incorporated radiofrequency hyperthermia into the tissue: comprehensive study with histology and pathology observations *IET Nanobiotechnol.* **13** 634–9

[45] Norahan M H, Amroon M, Ghahremanzadeh R, Rabiee N and Baheiraei N 2019 Reduced graphene oxide: osteogenic potential for bone tissue engineering *IET Nanobiotechnol*

[46] Willner I, Riklin A, Shoham B, Rivenzon D and Katz E 1993 Development of novel biosensor enzyme electrodes: glucose oxidase multilayer arrays immobilized onto self-assembled monolayers on electrodes *Adv. Mater.* **5** 912–5

[47] Preston G W and Wilson A J 2013 Photo-induced covalent cross-linking for the analysis of biomolecular interactions *Chem. Soc. Rev.* **42** 3289–301

[48] Karimi M, Mansouri M R, Rabiee N and Hamblin M R 2018 Advances in nature-inspired nanomaterials *Advances in Nanomaterials for Drug Delivery: Polymeric, Nanocarbon, and Bio-inspired* (San Rafael, CA: Morgan & Claypool)

[49] Karimi M, Mansouri M R, Rabiee N and Hamblin M R 2018 Aptamers and pathogen-based carriers *Advances in Nanomaterials for Drug Delivery: Polymeric, Nanocarbon, and Bio-inspired* (San Rafael, CA: Morgan & Claypool)

[50] Karimi M, Mansouri M R, Rabiee N and Hamblin M R 2018 Polymeric and hyper-branched nanoparticles and dendrimers *Advances in Nanomaterials for Drug Delivery: Polymeric, Nanocarbon, and Bio-inspired* (San Rafael, CA: Morgan & Claypool)

[51] Petach H, Ostroff R, Greef C and Husar G M 2004 Processing of photoaptamer microarrays *Methods Mol. Biol.* **264** 101–10

[52] Smyth G K 2005 Limma: linear models for microarray data *Bioinformatics and Computational Biology Solutions using R and Bioconductor* (Berlin: Springer) pp 397–420

[53] Brazma A *et al* 2001 Minimum information about a microarray experiment (MIAME)—toward standards for microarray data *Nat. Genet.* **29** 365

[54] Templin M F, Stoll D, Schrenk M, Traub P C, Vöhringer C F and Joos T O 2002 Protein microarray technology *Drug Discov. Today* **7** 815–22

[55] Fodor S P, Read J L, Pirrung M C, Stryer L, Lu A T and Solas D 1991 Light-directed, spatially addressable parallel chemical synthesis *Science* **251** 767

[56] Wilson D S and Nock S 2003 Recent developments in protein microarray technology *Angew. Chem. Int. Ed.* **42** 494–500

[57] Angenendt P 2005 Progress in protein and antibody microarray technology *Drug Discov. Today* **10** 503–11

[58] Gao X, LeProust E, Zhang H, Srivannavit O, Gulari E, Yu P, Nishiguchi C, Xiang Q and Zhou X 2001 A flexible light-directed DNA chip synthesis gated by deprotection using solution photogenerated acids *Nucl. Acids Res.* **29** 4744–50

[59] Li S, Bowerman D, Marthandan N, Klyza S, Luebke K J, Garner H R and Kodadek T 2004 Photolithographic synthesis of peptoids *J. Am. Chem. Soc.* **126** 4088–9

[60] Angenendt P, Glökler J, Sobek J, Lehrach H and Cahill D J 2003 Next generation of protein microarray support materials: evaluation for protein and antibody microarray applications *J. Chromatogr.* A **1009** 97–104

[61] López-Pérez P M, Grimsey E, Bourne L, Mikut R and Hilpert K 2017 Screening and optimizing antimicrobial peptides by using SPOT-synthesis *Front. Chem.* **5** 25

[62] Zander N 2004 New planar substrates for the *in situ* synthesis of peptide arrays *Mol. Divers.* **8** 189–95

[63] Dirk F H W 2011 Chemistry of SPOT synthesis for the preparation of peptide macroarrays on cellulose membranes *Mini-Rev. Org. Chem.* **8** 114–20

[64] Babu H K R R and Gheber L A 2018 Rapid assaying of miniaturized protein microarray *Sens. Actuators* B **268** 55–60

[65] Tomkins J E, Dihanich S, Beilina A, Ferrari R, Ilacqua N, Cookson M R, Lewis P A and Manzoni C 2018 Comparative protein interaction network analysis identifies shared and distinct functions for the human ROCO proteins *Proteomics* **18** 1700444

[66] Clancy K F, Dery S, Laforte V, Shetty P, Juncker D and Nicolau D V 2019 Protein microarray spots are modulated by patterning method, surface chemistry and processing conditions *Biosens. Bioelectron.* **130** 397–407

[67] Xiao Q, Wu J, Dang P and Ju H 2018 Multiplexed chemiluminescence imaging assay of protein biomarkers using DNA microarray with proximity binding-induced hybridization chain reaction amplification *Anal. Chim. Acta* **1032** 130–7

[68] Levine Z G, Fan C, Melicher M S, Orman M, Benjamin T and Walker S 2018 O-GlcNAc transferase recognizes protein substrates using an asparagine ladder in the tetratricopeptide repeat (TPR) superhelix *J. Am. Chem. Soc.* **140** 3510–3

[69] Falsey J R, Renil M, Park S, Li S and Lam K S 2001 Peptide and small molecule microarray for high throughput cell adhesion and functional assays *Bioconjug. Chem.* **12** 346–53

[70] Yousaf M N and Mrksich M 1999 Diels–Alder reaction for the selective immobilization of protein to electroactive self-assembled monolayers *J. Am. Chem. Soc.* **121** 4286–7

[71] Inwood S, Buehler E, Betenbaugh M, Lal M and Shiloach J 2018 Identifying HIPK1 as target of miR-22-3p enhancing recombinant protein production from HEK 293 cell by using microarray and HTP siRNA screen *Biotechnol. J.* **13** 1700342

[72] Houseman B T, Gawalt E S and Mrksich M 2003 Maleimide-functionalized self-assembled monolayers for the preparation of peptide and carbohydrate biochips *Langmuir* **19** 1522–31

[73] Reddy M M and Kodadek T 2005 Protein 'fingerprinting' in complex mixtures with peptoid microarrays *Proc. Natl Acad. Sci.* **102** 12672–7

[74] Lei L, Xie J, Yu J, Li Y and Liu Y 2018 Parallel study on protein O-GlcNAcylation in prostate cancer cell with a sensitive microarray biochip *Anal. Biochem.* **558** 53–9

[75] Sun X, Wang H, Wang Y, Gui T, Wang K and Gao C 2018 Creation of antifouling microarrays by photopolymerization of zwitterionic compounds for protein assay and cell patterning *Biosens. Bioelectron.* **102** 63–9

[76] Salisbury C M, Maly D J and Ellman J A 2002 Peptide microarrays for the determination of protease substrate specificity *J. Am. Chem. Soc.* **124** 14868–70

[77] Zhu Q, Uttamchandani M, Li D, Lesaicherre M L and Yao S Q 2003 Enzymatic profiling system in a small-molecule microarray *Org. Lett.* **5** 1257–60

[78] Goddard J-P and Reymond J-L 2004 Enzyme activity fingerprinting with substrate cocktails *J. Am. Chem. Soc.* **126** 11116–7

[79] Kiyonaka S, Sada K, Yoshimura I, Shinkai S, Kato N and Hamachi I 2004 Semi-wet peptide/protein array using supramolecular hydrogel *Nat. Mater.* **3** 58–64

[80] Cheng C-W, Lin K-C, Pan F-M, Sinchaikul S, Wong C-H, Su W C, Hsu C-H and Chen S-T 2004 Facile synthesis of metal-chelating peptides on chip for protein array *Bioorg. Med. Chem. Lett.* **14** 1987–90

[81] Chiari M, Cretich M, Corti A, Damin F, Pirri G and Longhi R 2005 Peptide microarrays for the characterization of antigenic regions of human chromogranin A *Proteomics* **5** 3600–3

[82] Andresen H, Grotzinger C, Zarse K, Kreuzer O J, Ehrentreich-Forster E and Bier F F 2006 Functional peptide microarrays for specific and sensitive antibody diagnostics *Proteomics* **6** 1376–84

[83] Yu X, Petritis B, Duan H, Xu D and LaBaer J 2018 Advances in cell-free protein array methods *Expert Rev. Proteomics* **15** 1–11

Chapter 4

Protein microarrays: conclusions and future perspectives

Navid Rabiee, Sakineh Hajebi, Sepideh Ahmadi, Mohammad Rabiee, Mojtaba Bagherzadeh, Saeid Maghsoudi and Michael R Hamblin

Microarray technology makes possible the simultaneous investigation of many biological interactions and has been widely used in the fields of genomics (study of the set of genes within an individual) and proteomics (study of the set of proteins within an individual). This method has very high efficiency and is able to analyze a considerable amount of information in a short period of time. Proteins are the workhorses of the cell and play vital roles in both healthy cells and cells from disease states In order to understand how proteins interact with each other, and how they govern different cellular processes it is important to explore their behavior and interactions under a wide variety of experimental conditions. Unlike the genome, which includes a constant number of genes, the proteome of different cells is extremely dynamic. Proteins are continuously processed inside the body in response to external stimuli and undergo a wide variety of post-translational modifications. Consequently, it is difficult to reliably determine the exact number or quantities of proteins that exist within a human, animal, or other biological system at any specific time. Furthermore, protein families are very variable, and proteins have significant differences in their physical size, affinity constants, chemical and structural properties, and their relative abundance inside the cells. As a result, it is very challenging to accurately describe all such protein–protein interactions.

Protein microarrays are widely used in proteomic studies and a variety of purposes. High-throughput processing has become a tendency due to high productivity and cost effectiveness. Given a high-throughput system, protein microarrays accelerate new discoveries in protein interaction for clinical investigation. One of the advantages of technology over other techniques, is the reduction in sample usage. This feature can be important in clinical trials using rare samples, such as human

doi:10.1088/978-0-7503-3667-3ch4

serum samples, and may increase the sensitivity and potency of protein microarrays. To achieve uniform reproducibility of standard methods and decrease in the production costs, this technology will be a favorable tool for future use.

In general, the application of protein microarrays can be discussed in two further areas: in high-throughput analysis and clinical diagnosis. High-throughput analysis means measuring thousands of protein–protein interactions that can rapidly identify candidate drugs and their mechanisms of action. Numerous diagnostic protein microarrays have been approved by the Food and Drug Administration (FDA). The diagnosis of autoimmune and infection diseases is the attention of greatest of the protein microarrays such as the Bio-Plex™ 2200 system (https://www.bio-rad.com/).

Combining protein microarray with other techniques such as the enzyme-linked immunosorbent assay (ELISA) and lateral flow assay (LFA) can make routine the use of this technique in clinics and laboratories. The LFA technique has advantages such as fast and direct detection (colorimetric detection), multi-analyte diagnostic and no need for expensive equipment. Combining this technique with protein microarray can increase the sensitivity and lower sample consumption [1–3].

Protein microarrays such as DNA microarray technology can be used to describe expression profiles of specific disease states. Antibody-based or functional based protein array against specific biomarkers can be used in the production of personal biosensors in the future. No change in the structure of the proteins used to immobilize, has led to development of easier and more effective functional assays than conventional ones. On the other hand, manufacture of appropriate capture molecules should be considered. It is generally difficult to find suitable antibodies to bind to intracellular proteins, one of the most important cases is high affinity and low dissociation constant for antibody. The dissociation constant in classical antibodies is 10^{-9} to 10^{-12} M, however, expensive production and time consumption are some of its major disadvantages. Protein validation is another example; nowadays about 30 000 000 antibodies can be purchased that cover about 95% of human genes, but all of this cannot be used in presentation technology (Anitbodypedia: http://www.antibodypedia.org).

Therefore, it is essential to find some methods to produce cheaper antibodies or to replace some other protein-binding compounds as capture molecules, such as nucleic acids that are easier to synthesize and are a great source of protein absorption. This mRNA–protein fusion, as antibody mimics, have been displayed to have dissociation constants in the range 10^{-9} to 10^{-12} M [4, 5]. DNA Microarrays are now widely used in diagnosis and clinical research, and as a complement to other techniques in genotype detection as well as to measure transcriptional activity of genes. But with the increase in the ease of ease of their manufacture, protein microarray assays will soon be applied to discovery of the protein interactions and biomarkers.

An array is a piece of silicon, glass or other material onto which diverse protein or peptide molecules, or other particular DNA binding sequences can be attached at distinct locations in a spatially ordered manner, therefore forming a microscopic array [6–9].

Most of the previous discussion has concentrated on two-dimensional B–D surface constructions. However, sensitive proteins may not be able to remain intact in the harsh conditions typical of 2D surfaces, because the functional properties of a particular protein are highly dependent on its structure and tertiary conformation. As a consequence, other designs have been proposed based on C–D surfaces in three dimensions with a fluidic environment that more closely mimics the biological surroundings of natural proteins. Hydrogels and polymer membranes are some examples of 3D-based systems. The main advantage of the 3D layout is that the conformation and activity of the protein are preserved by the fluidic environment. The higher probe density and larger surface area are other advantages of the 3D architecture. The first substrates used for the preparation of microarrays were polymer membranes such as nitro-cellulose [10, 11] or aminated nylon [12]. For more than two decades, microbiology laboratories extensively employed these platforms as blotting supports, using radioimaging and other recognition technologies such as fluorescence to detect proteins on these supports. Most of the studies on membrane supports have employed downstream methods such as mass spectrometry in proteomic investigations. The special benefits of these platforms are related to sample preparation and overall cost, because antibodies directly bind to proteins on the blotting membrane surface after electrophoresis.

Biacore has pioneered the extensive use of SPR technology in protein or peptide arrays constructed in hydrogel-modified gold substrates, by taking advantage of the high increases in sensitivity provided by the surface plasmons, to advance the field of real-time protein interaction studies over the last few years. Polymer hydrogels and polymer brushes on glass substrates in traditional microarray arrangements have also been recently expanded, however, their usage is still restricted [13–18].

Over the past few years, various studies have demonstrated the success of protein microarray technology in the detection of various diseases, biomarkers, and drug discovery. However, there is still a need to study and develop cheaper platforms, commercialization, and a comprehensive framework for data analysis. The large-scale production of high-quality antibodies is also due to the complexity of expression and purification methods and the lack of labeling of the samples, which are still limitations. As a result, there is a need to develop systems that can be used without labeling the sample. Meeting the challenges ahead requires more study and effort. Currently, technologies such as next-generation sequencing (NGS) and the like, can fill the gap in protein microarray. As a result, more efforts need to be made to develop ideas and tools to improve this technology, so that this technology can achieve a real potential as a robust platform for clinical and proteomics research.

References

[1] Waltari E, Carabajal E, Sanyal M, Friedland N and McCutcheon K M 2019 Adaption of a conventional ELISA to a 96-well ELISA-array for measuring the antibody responses to influenza virus proteins, viruses and vaccines *J. Immunol. Methods* **481–482** 112789 2019.12.20.885285

[2] Lama L 2017 Novel methods for improving rapid paper-based protein assays with gold nanoparticle detection *Licentiate thesis, monograph* (Stockholm: KTH Royal Institute of Technology)

[3] Chinnasamy T, Segerink L I, Nystrand M, Gantelius J and Svahn H A 2014 A lateral flow paper microarray for rapid allergy point of care diagnostics *Analyst* **139** 2348–54

[4] Weng S, Gu K, Hammond P W, Lohse P, Rise C, Wagner R W, Wright M C and Kuimelis R G 2002 Generating addressable protein microarrays with PROfusion covalent mRNA-protein fusion technology *Proteomics* **2** 48–57

[5] Xu L *et al* 2002 Directed evolution of high-affinity antibody mimics using mRNA display *Chem. Biol.* **9** 933–42

[6] Ahmadi S, Rabiee N and Rabiee M 2019 Application of aptamer-based hybrid molecules in early diagnosis and treatment of diabetes mellitus: from the concepts towards the future *Curr. Diabetes Rev.* **15** 309–13

[7] Noorirad S N, Pourghasem M, Feizi F, Abedian Z, Ghasemi M, Babazadeh Z and Rabiee N 2018 Time dependent of epigenetic effect of disulfiram on tumor suppressor gene of RASSF1A in Hela cancer cell line *Basic Res. Med. Sci.* **5** 8–13

[8] Rabiee N, Bagherzadeh M and Rabiee M 2019 A perspective to the correlation between Brain insulin resistance and Alzheimer: medicinal chemistry approach *Curr. Diabetes Rev.* **15** 255–8

[9] Hajebi S, Rabiee N, Bagherzadeh M, Ahmadi S, Rabiee M, Roghani-Mamaqani H, Tahiri M and Hamblin M R 2019 Stimulus-responsive polymeric nanogels as smart drug delivery systems *Acta Biomater.* **92** 1–18

[10] Fici D A, McCormick W, Brown D W, Herrmann J E, Kumar V and Awdeh Z L 2010 A protein multiplex microarray substrate with high sensitivity and specificity *J. Immunol. Methods* **363** 60–6

[11] Towbin H, Staehelin T and Gordon J 1979 Electrophoretic transfer of proteins from polyacrylamide gels to nitrocellulose sheets: procedure and some applications *Proc. Natl. Acad. Sci.* **76** 4350–4

[12] Gershoni J M and Palade G E 1982 Electrophoretic transfer of proteins from sodium dodecyl sulfate-polyacrylamide gels to a positively charged membrane filter *Anal. Biochem.* **124** 396–405

[13] Karimi M, Mansouri M R, Rabiee N and Hamblin M R 2018 Drug delivery approaches *Advances in Nanomaterials for Drug Delivery: Polymeric, Nanocarbon, and Bio-inspired* vol 1 (San Rafael, CA: Morgan & Claypool)

[14] Karimi M, Mansouri M R, Rabiee N and Hamblin M R 2018 Carbon-based nanomaterials *Advances in Nanomaterials for Drug Delivery* (San Rafael, CA: Morgan & Claypool)

[15] Rabiee N, Kiani M, Bagherzadeh M, Rabiee M and Ahmadi S 2019 *Nanoparticle (NP)-Based Delivery Vehicles* (San Rafael, CA: Morgan & Claypool)

[16] Rabiee N, Rabiee M, Bagherzadeh M and Hamblin M R 2019 *Stimuli-Responsive Polymers: Nano-Dimension* (San Rafael, CA: Morgan & Claypool)

[17] Rabiee M, Rabiee N, Salarian R and Rabiee G 2019 *Introduction to Nanomaterials in Medicine* (San Rafael, CA: Morgan & Claypool)

[18] Rabiee N, Kiani M, Bagherzadeh M, Rabiee M and Ahmadi S 2019 Aptamer-based nanostructures *Nanoparticle (NP)-Based Delivery Vehicles* (San Rafael, CA: Morgan & Claypool)

Lightning Source UK Ltd.
Milton Keynes UK
UKHW051315091220
374844UK00002B/41

9 780750 336659